Water Governance and Civil Society Responses in South Asia

This cluster of books presents innovative and nuanced knowledge on water resources, based on detailed case studies from South Asia—India, Bangladesh, Bhutan, Nepal, Pakistan and Sri Lanka. In providing comprehensive analyses of the existing economic, demographic and ideological contexts in which water policies are framed and implemented, the volumes argue for alternative, informed and integrated approaches towards efficient management and equitable distribution of water. These also explore the globalization of water governance in the region, particularly in relation to new paradigms of neoliberalism, civil society participation, integrated water resource management (IWRM), public–private partnerships, privatization and gender mainstreaming.

Water Resources Policies in South Asia
Editors: Anjal Prakash, Sreoshi Singh, Chanda Gurung Goodrich and S. Janakarajan
ISBN 978-0-415-81198-9

Globalization of Water Governance in South Asia
Editors: Vishal Narain, Chanda Gurung Goodrich, Jayati Chourey and Anjal Prakash
ISBN 978-0-415-71066-4

Informing Water Policies in South Asia
Editors: Anjal Prakash, Chanda Gurung Goodrich and Sreoshi Singh
ISBN 978-0-415-71059-6

Water Governance and Civil Society Responses in South Asia
Editors: N. C. Narayanan, S. Parasuraman and Rajindra Ariyabandu
ISBN 978-0-415-71061-9

Water Governance and Civil Society Responses in South Asia

Editors
N. C. Narayanan
S. Parasuraman
Rajindra Ariyabandu

First published 2014 in India
by Routledge
912 Tolstoy House, 15–17 Tolstoy Marg, Connaught Place, New Delhi 110 001

Simultaneously published in the UK
by Routledge
2 Park Square, Milton Park, Abingdon, Oxon, OX14 4RN

Routledge is an imprint of the Taylor & Francis Group, an informa business

© 2014 South Asia Consortium for Interdisciplinary Water Resources Studies (SaciWATERs)

Typeset by
Glyph Graphics Private Limited
23, Khosla Complex
Vasundhara Enclave
Delhi 110 096

All rights reserved. No part of this book may be reproduced or utilized in any form or by any electronic, mechanical or other means, now known or hereafter invented, including photocopying and recording, or in any information storage and retrieval system without permission in writing from the publishers.

British Library Cataloguing-in-Publication Data
A catalogue record of this book is available from the British Library

ISBN 978-0-415-71061-9

Contents

List of Tables   vii
List of Figures   ix
List of Abbreviations   xi
Foreword by Peter P. Mollinga   xvii
Acknowledgements   xix

Introduction: Paradigm Shifts in Water Governance
and Civil Society Responses in South Asia: A Review   1
N.C. Narayanan, S. Parasuraman and
Rajindra Ariyabandu

1. The State, Civil Society and Challenges
   to Democratic Governance in South Asia   31
   Mahfuzul H. Chowdhury

Part I: NGOs in Collaborative Advocacy

2. Watershed Development Policies and Programs
   in India: Scope and Constraints of Civil Society Action   55
   K. J. Joy

3. How an NGO Works the State and Why it Succeeds:
   A Case Study from Central India   90
   Vasudha Chhotray

Part II: Foreign Assistance and Water Governance

4. Water Provision and Modernity:
   The Consequence of Foreign Aid in Nepal   123
   Sudhindra Sharma

5. Foreign Assistance, Dependence and Debt:
   Sanitation Case Study, Kandy, Sri Lanka 144
   Sunil Thrikawala and *N.C. Narayanan*

6. Institutional Changes, Public Provision
   and Drinking Water Supply in Kerala 170
   N. C. Narayanan and *S. Mohammed Irshad*

Part III: NGOs in Policy Influence

7. Water Policy Development in a Multi-party
   System of Governance: A Case Study of Sri Lanka 201
   Rajindra Ariyabandu

8. Water Rights in Civil Society and Governance:
   The Sri Lankan Experience 236
   Ruana Rajepakse

9. "Flood Action Plan" and NGO Protests
   in Bangladesh: An Assessment 263
   Hamidul Huq

Part IV: Social Movements in Water Governance

10. The Water Rights Movement in South
    Maharashtra, India 285
    Suhas Paranjape and *Seema Kulkarni*

11. Lessons from Plachimada: From Anti-Coca-Cola
    Agitation to Democratization of Water 309
    C. R. Bijoy

About the Editors 343
Notes on Contributors 344
Index 347

List of Tables

4.1	Allocations for drinking water supply in various five-year plans in Nepal	129
5.1	Government debt indicators in Sri Lanka	148
5.2	Operational expenditure of drinking water over the years in Sri Lankan rupees	150
5.3	Piped water production and non-revenue water over time (all SLR)	152
5.4	Non-revenue water and costs incurred	152
5.5	Types of non-revenue water	153
5.6	Water treatment costs	155
5.7	Estimated costs of the sewage treatment plant	158
5.8	Direct operation and maintenance costs of kandy city water supply augmentation and environmental improvement project	163
6.1	Production, distribution and leakage of piped water in Kerala (1999–2006)	177
6.2	Category-wise arrears of water charges in Kerala	179
6.3	Estimated cost and loan part of Japanese bank of international cooperation-aided drinking water project	183
6.4	Share of contributions to jalanidhi (INR, in millions)	188
6.5	Average monthly expenditure of beneficiary groups	189

List of Figures

5.1	Total revenue, development expenditure and loan interest over the years	151
5.2	Fund Utilization by NWSDB	156
5.3	Kandy city water supply augmentation and environmental improvement project and its implications	159
6.1	Sources of funding drinking water in Kerala	174
6.2	Financial position of KWA	175
6.3	Capital expenditure on urban and rural water supply projects	176
6.4	KWA's income from water supply and total revenue receipts	178
6.5	Plan and non-plan expenditure of KWA, 1996–2006	179
6.6	Ratio of government of Kerala grant to domestic capital investment of KWA	180
6.7	Foreign aid and government of Kerala funding for water sector	181
6.8	Organizational chart of Japanese Bank of international cooperation-funded project	185

List of Abbreviations

ACC	Anti-Corruption Commission
ADAB	Association of Development Agencies in Bangladesh
ADB	Asian Development Bank
ATREE	Ashoka Trust for Research in Ecology and the Environment
BAIF	Bharatiya Agro Industries Foundation
BOD	Biochemical Oxygen Demand
BWDB	Bangladesh Water Development Board
CAPART	Council for Promotion of Participatory Action and Rural Technology
CARE	Cooperative for American Relief Everywhere
CASAD	Centre for Applied Systems Analysis in Development
CBO	Community Based Organization
CEA	Central Environmental Authority
CEN	Coalition of Environmental NGOs
CIDA	Canadian International Development Agency
CISED	Centre for Interdisciplinary Studies in Environment and Development
CMP	Common Minimum Programme
CPI	Communist Party of India
CPI(M)	Communist Party of India (Marxist)
CPP	Compartmentalisation Pilot Project
CSE	Centre for Science and Environment
CSO	Civil Society Organization
CWRMP	Comprehensive Water Resource Management Project
CWS	Community Water Supply
DANIDA	Danish International Development Agency
DFID	Department For International Development
DISVI	Associazione Internazionale Disarmo e Sviluppo
DPAP	Drought Prone Area Programme
DWSS	Department of Water Supply and Sewerage

EGS	Employment Guarantee Scheme
EIA	Environmental Impact Assessment
FAO	Food and Agriculture Organisation
FAP	Flood Action Plan
FCRA	Foreign Currency Regulation Act
FERA	Foreign Exchange Regulation Act
FO	Farmers' Organisation
ForWaRD	Forum for Watershed Research and Policy Dialogue
FPCO	Flood Plan Coordination Organisation
GDP	Gross Domestic Product
GIDR	Gujarat Institute of Development Research
GO	Government Organization
GoI	Government of India
GONGO	Government Organised Non-Governmental Organization
GP	Gram Panchayats
GRO	Grass-Root Organisation
GSO	Grassroot Support Organisation
GTZ	German Agency for Technical Cooperation
Ha	Hectare
HCCB	Hindustan Coca-Cola Beverages
IAO	Information, Alternatives and Opposition
ICAR	Indian Council of Agricultural Research
ICSECR	International Covenant on Social, Economic and Cultural Rights
IEE	Initial Environmental Examination
IMPSA	Irrigation Management Policy Support Activity
IRA	Independent Water Resources Regulatory Authority
IRN	International River Network
IWMI	International Water Management Institution
IWRM	Integrated Water Resources Management
JBIC	Japan Bank for International Cooperation
JICA	Japan International Cooperation Agency
KAWAD	Karnataka Watershed Development Society
KCWSAEIP	Kandy City Water Supply Augmentation and Environmental Improvement Project

KMC	Kandy Municipal Council
KRWSA	Kerala Rural Water and Sanitation Agency
KSPCB	Kerala State Pollution Control Board
KWA	Kerala Water Authority
LEISA	Low External Input-Based Sustainable Agriculture
LGED	Local Government Engineering Department
LNP	Lal Nishan Party
LSD	Local Self-Government Department
LVS	Lok Vidnyan Sanghatana
MALLI	Ministry of Agriculture, Lands, Livestock and Irrigation
MASL	Mahaweli Authority of Sri Lanka
MIWRM	Ministry of Irrigation and Water Resources Management
MKVDC	Maharashtra Krishna Valley Development Corporation
MLA	Member of Legislative Assembly
MMRBRD	Ministry of Mahaweli, River Basin and Rajarata Development
MNC	Multi-National Corporation
MoRD	Ministry of Rural Development
MPLD	Ministry of Panchayat and Local Development
MSM	Mukti Sangharsh Movement
MYRADA	Mysore Resettlement and Development Agency
NAPM	National Alliance of People's Movements
NBA	Narmada Bachao Andolan
NEWAH	Nepal Water for Health
NGO	non-governmental organization
NRMS	Natural Resources Management Services Private Limited
NWDPRA	National Watershed Development Project for Rainfed Areas
NWP	National Water Policy
NWRA	National Water Resources Authority
NWRC	National Water Resources Management Council
NWSDB	National Water Supply and Drainage Board
O & M	Operation and Management

OECF	Overseas Economic Cooperation Fund
PAP	Parambikulam-Aliyar Project
PHED	Public Health Engineering Department
PIA	Project Implementation Agency
PIM	Participatory Irrigation Management
PM	Prime Minister
PPP	Public Private Partnership
PRI	Panchayati Raj Institution
PVO	Private Voluntary Organization
PWD	Public Works Department
RBC	River Basin Committees
RBO	River Basin Organization
RTI	Right To Information
RWRA	Regional Water Resources Agencies
RWRC	Regional Water Resources Councils
SC	Scheduled Caste
SGO	Semi-Government Organization
SHG	Self Help Group
SKSS	Shetmajoor Kashtakari Shetkari Sanghatana
SMD	Shramik Mukti Dal
SMM	South Maharashtra Movement
SO	Support Organization
SOPPECOM	Society for Promoting Participative Ecosystem Management
SPS	Samaj Pragati Sahyog
SSP	Sardar Sarovar Project
ST	Scheduled Tribes
STA	Senior Technical Advisor
STP	Sewage Treatment Plant
TEC	Tokyo Engineering Company
TIB	Transparency International Bangladesh
TMC	Thousand Million Cubic Feet
TRIPS	Trade Related Intellectual Property Rights
UN	United Nations
UNDP	United Nations Development Programme
UNICEF	United Nations Children's Fund
USA	United States of America
USAID	United States Agency for International Development

USD	United States Dollar
WARASA	Watershed Areas' Rainfed Agricultural Systems Approach
WASSAN	Watershed Support Services and Activities Network
WHO	World Health Organization
WIRP	Wasteland Integrated Research Programme
WMB	Water Mediation Board
WMIP	Water Management Improvement Project
WRC	Water Resources Council
WRMP	Water Resources Management Project
WRS	Water Resources Secretariat
WRT	Water Resources Tribunal
WTO	World Trade Organization
WUA	Water Users' Associations

Foreword

Water Governance and Civil Society Responses in South Asia is part of the series of 'Readers' developed by the SaciWATERs. The series of readers on topical water resource management studies is developed by the 'Crossing Boundaries' project which focuses on regional capacity building on IWRM, Gender and Water in South Asia. The series is a collective effort with each volume being edited by two or three editors and consists of contributions from scholars and professionals representing South Asian sub-regional countries. The series seeks to provide high quality research contribution on different water resources topics accessible to a broader interested audience and suitable as a resource material for education and capacity building.

This volume on water governance attempts to bring together contributions from researchers and practitioners in the water sector to examine the strategic role of the Civil Society Organizations (CSOs) and Social Movements outside the conventional sphere of the State and business. The central question this volume attempts to address is the paradigm shifts in aspects of water policy/governance and the role of the CSOs in influencing public policy. This becomes relevant since most of the contemporary governance debates over water, centers around either institutional reforms within the State apparatus or the scope and constraints to neo-liberal shift to market mechanisms. The goal of the civil society initiatives is to ensure democratic governance and access and control of water by the people including the poor and the marginalized communities.

The objectives of this reader are of three-fold; *a)* to understand the paradigm shift in water policy and governance in South Asia over the past three decades, *b)* to understand the role of the civil society organizations/NGOs within the sphere of civil society in addressing these changes, especially the scope and constraints of influencing public policy either in collaboration or in confrontation with the State, *c)* to map possible implications of these changes for concerns like social equity and democratic participation. The collection of papers in this reader will bring out the influence, effectiveness, drawbacks and challenges faced by CSOs and voluntary organizations and social movements at different levels in South Asian countries. The volume addresses four key areas of interest; Water Governance and Civil Society, Policy Shifts, Donor Influence and Policy and NGOs and Social Movements.

Trust that the information and insights provided in this volume will inspire a new generation of water professionals to think beyond the conventional technical focus of water resources education and incorporate important concerns like equity and democratic participation and encourage an analysis of water-politics, role of the civil society and its implications on water governance.

Peter P. Mollinga
Professor, Department of Development Studies
School of Oriental and African Studies
University of London, UK

Acknowledgements

This book is the effort and contribution of a number of people from the initial stages of a workshop to this book. Peter Mollinga and Nimal Gunawardena encouraged us to such an idea of a book. Anju Helen George helped us organise the workshop that discussed the papers in Pune in 2007. Professor D.N. Dhanagare initially reviewed the articles and provided invaluable comments. Dibya Ratnakanasakar, Anjal Prakash and other staff of SaciWATERs constantly followed up the manuscripts for publication. Suparna Sengupta and Denise Fernandes assisted at various stages of work. All the contributors obliged us by readily verifying the changes and suggestions. We wish to acknowledge the support received from the editorial team at Routledge India. The book would not have been possible without the financial support of the Netherlands government through the 'Crossing Boundaries Project' through the Irrigation and Water Engineering Department, Wageningen University, The Netherlands.

Our gratitude to all the above mentioned is beyond words.

Introduction

Paradigm Shifts in Water Governance and Civil Society Responses in South Asia
A Review

N.C. Narayanan, S. Parasuraman and
Rajindra Ariyabandu

Introduction

Water governance in South Asia is witnessing profound changes and this volume explores the related paradigm shifts during the last three decades. The focus is on social equity and democratic participation, two stated aims of water sector governance reforms and the accomplishments along these lines. The contributors are researchers and activist scholars working in water sector development in the region and a core issue is the role of the civil society organizations (CSOs) in the shifts in water governance. The volume illustrates many interesting developments in policy formulation,[1] donor-State nexus, CSO/non-governmental organization (NGO) interventions and civil society thrust on safeguarding fundamental rights to water. The major thrust and the common thread throughout this volume reflects the implications of the paradigm shift in policy and governance in water resources management and its responses in contemporary South Asia.

As a background, this chapter reviews the governance of water in South Asia with a focus on the paradigm shifts, which relates to the thrust on principles of cost recovery, decentralization and participation (not only of people but the spheres of civil society and market institutions) in water governance mediated by the international financial institutions and donor agencies in the last three decades. Some of the key issues raised in this context are as follows – (*a*) How has the role of the state changed over the past three decades? (*b*) What are the guiding principles in the water sector and what is its impact on the institutions of water governance in South Asia? (*c*) What are the implications of these changes for social equity and democratic participation and what are the responses?

The chapter begins with a discussion of the evolution of the conventional state-led water governance, which is a colonial legacy of water management by command and control of the State with a thrust on "productive" use of water. The limits to this approach and challenges from the angles of environmental and social externalities have triggered alternatives that are more decentralized, participatory and locally managed. The first section critically examines both the conventional water governance in South Asia and certain suggested alternatives. The second section reviews the paradigm shifts in water governance starting with the triggers of shift, new reigning principles in the current strategies, programmatic interventions in key sub sectors like irrigation and drinking water and concludes with a review of the implications of the paradigm shift. The third section presents the concept and discussions on civil society briefly along with the responses to the changes in water governance and outlines the arena of responses through the contributions in this volume.

State-led Governance, Limits and Alternatives

Large Scale Storages and Externalities

The conventional command and control based state-led water governance in the past century is typically reflected in the construction of large irrigation systems and transmission of water to enhance the productive uses of water. The history of colonial policies in South Asia was one of maximizing revenue through irrigation projects which accrued one of the highest rates of return to investment. The post colonial governments carried forward the legacy and along with private initiatives enhanced irrigation development (Shah 2009). Although the colonial approach to building dams to prevent famines (Atwood 2007) has been contested by later studies (Shah 2009), the thrust on agricultural modernization ensured a central role to big storages. The Green Revolution-inspired strategies for agriculture development in post-independent India coincided with the large irrigation projects to supply assured water that was a crucial input to the envisioned agricultural growth.[1] A major criticism to such a development imagination was its extraneous nature, especially in the era of "development assistance" and "technology transfer" to help the newly independent ex-colonies of a fast pace of development by modernisation. The construction of large dams and other infrastructure for water management was actively supported by international financial institutions and many donors until the 1990s. Water projects were not only dams, but other engineering structures like embankments to ward-off floods and increase the cropping intensity. In 1964, the Government of Pakistan, drawing upon the Krug Mission recommendations, charted out a 20-year Master Plan to build thousands of kilometers of embankments in present day Bangladesh by hiring foreign and local consultants to develop the Flood Action Plan. This was supported by the United States Agency for International Development (USAID) for the Green Revolution program that attempted to double the triple crop extent, irrigate and protect cultivable lands by flood control and drainage schemes (Hossain 1994). The justification for such investments was agricultural growth, food security, increase in employment, decline in rural poverty, etc. The best documentation of the related perspectives and practices are from a World Bank expert who had the privilege to work on financing water infrastructure in the three South Asian countries.[2]

In the Indian case, according to the Constitution of India, water is a state subject with states exercising power over water storage, supplies, rivers, hydropower, irrigation canals, fisheries, etc. and thus owning enormous state-owned infrastructure for water management.[3] In the federal system, the Union government's powers in water are restricted to shipping and navigation, national waterways, tidal and territorial waters and inter-state water disputes.[4] The criticisms to the State-led water infrastructure management is the bundling of governance tasks like policy making, legislating, implementing and regulating vested with the government, which is likely to result in inefficiency and wastage due to the lack of transparency and accountability of the state system. The typical elements of such state failure as suggested by a World Bank led study are as follows – poor infrastructure, incomplete projects, insufficient funds due to poor collection efficiency leading to lack of operation and maintenance and thus, a vicious cycle of inefficiency. This leads to conflicts between uses and the failure of public provisioning of water leads to increased dependence on ground water triggering a cycle of overuse and environmental degradation (Briscoe and Malik 2006). This efficiency-based criticism is marginal compared to a different set of criticisms hinged on the social and environmental externalities of large water storages (Bandyopadhyay 2002; Cernea 1999; Patankar and Phadke 2008; World Bank 1998).

Water management to enhance the productive uses like irrigated agriculture and industry change water regimes, with consequences for the distribution and allocation of water resources among different stakeholders (Chowdhury et al. 1997). Although the intention of water resource development projects is to provide economic benefits to society, the benefits and costs may not be distributed equally to all sections of society (Patankar and Phadke 2006; Sangameswaran 2009). A region's hydraulic endowments to already empowered beneficiaries with the costs disproportionately borne by dispossessed and marginal communities are downplayed (D'souza 2008). The classic example is that of the controversial Sardar Sarovar Project, where different cost-benefit outcomes were arrived due to political influence rather than being based on objective criteria that led to conflicts (Dwivedi 2006). Finally, large-scale water infrastructure is suggested over alternatives because of the strong triangle of interests and power of bureaucrats, politicians and engineers that favor them.

Emerging Alternatives

The equity rationale over efficiency has provided the basis for water allocation in a number of South Asian countries. The NGO Proshika's efforts in Bangladesh, the Andhi Khola Irrigation

Scheme in Nepal and the Sukhomajri and Pani Panchayat experiments in northern and western India respectively, all of which gave provision of water rights ensuring that the endowment is converted to entitlements, respectively (Apte 2001; Patil 2000; Sangameswaran 2009). More recently the pro-public judiciary in Sri Lanka has upheld the fundamental rights of its citizens to rights to water, to occupation and to prevent pollution of water resources (Rajepakse in this volume). Thus the projected water crisis in South Asia is suggested to be one of mismanagement more than scarcity, which calls for a shift from supply-side engineering to restraining the increase in demand, for conservation and more equitable management (Iyer 2007). Watershed development seemed one of the probable answers to the livelihood crisis and degradation for areas that have otherwise been bypassed by economic development, typically in dry lands, semi-arid and arid regions, perpetually under the shadow of drought (Shah et al. 1998). These experiments of locally managed systems in rain-fed areas also provided a direction for policy reforms from the mid-1990s to focus on greater allocation for the development of rain-fed areas through watershed management programs (GOI 2006). It was suggested that the focus needs to shift from big water resources development initiatives to primary and local water harvesting structures and watershed development programs. Once this approach is considered it would entail several changes in the design of governance of the water sector (Iyer 2007).

The emergence, consolidation and scaling-up of watershed development programmes from 1980 onwards is described in detail by K. J. Joy in Chapter Two of this volume. This shift acknowledged the needs of poorer sections of the population living in marginal lands and extended programmatic support for attaining the triple bottom line of environmental conservation, productivity enhancement, and greater inclusion (Kerr 2002). Importantly, the emphasis was the creation of watershed development committees with apparent total inclusion of the local community. However, inequalities on the basis of class, caste, patriarchy and ethnicity were identified as barriers to inclusion (Shah 2001) along with locational inequalities, primarily because of the biophysical characteristics of the watershed (Joy and Paranjape 2004). An attempt to understand the alternatives in the Rajasthan context (Narayanan and Kamath, 2012) revealed two types of shifts – first, small storages and watershed management and the second, more decentralized and private management of water. It is clear that both these shifts imply enhanced claims for availability of water at the local level and increased local influence in decision-making. However, such advantages do not ensure social equity and, in fact, was found to deepen existing inequalities in the light of democratic deficit in the formation of institutions of water management and governance (including Panchayati Raj institutions), which obviously is linked to the requisite of democratization needed in the larger political economy (ibid).

Beyond the traditional debate of big and small ways of managing water, the linkages of water governance to the large-scale changes happening in the macroeconomic sphere would be insightful. The neoliberal turn brought colossal changes in the global economy and at the level of national governments.[5] Following the oil shock, the mounting debt burden, structural adjustment and stabilization programs during the 1980s prompted contraction and austerity in most of the Third World (Watts 1994) that led to structural adjustment programs in national economies. With neoliberalism, development thinking has come back to square one, which prescribed one universal model for the whole world, partly due to the end of the Cold War which weakened the ideological and political backing for dissent. The next section outlines the trends of water sector reforms that emerged in South Asia in this macroeconomic context.

Paradigm Shifts

The paradigm shifts in the water sector in South Asia are part of the perspective changes globally that led to a comprehensive process of reforms in the sector. It started in the early 1990s partly by explicit policy statements and partly by programmatic shifts, both brought through the conditionalities attached to donor funding. The first two sub-sections will examine the principles underlying the reforms, the institutional drivers and changes and its implications. The process of reforms is based on and guided by global perspectives like Integrated Water Resource Management (IWRM) that emerged a decade back and gave the conceptual basis for the reform agenda.

Water Reforms Process

For the last two decades in South Asia, multilateral institutions with their donor lever suggested that the state lacked long-term plans and insisted that future lending would be conditional on acceptance of major policy changes. Correspondingly the move has been from supply side water management towards demand side water management (Sangameswaran 2010). The basic premise for doing so is the perceived inability of the government to effectively administer the existing water infrastructure, provide water in an economically efficient manner to users and muster further financial resources to infuse in the water infrastructure (Cullet 2006). About the rationale of such reforms, there are differing points that view these either as a response to the failure of the welfare state and of top-down technocratic approaches that needed change or a more critical approach that see the reforms as a direct outcome of the pressures of international financial institutions and of the influence of neoliberalism, or as involving the changing role of elites (Sangameswaran 2010). A mixture of internal factors like water scarcity, performance deterioration and financial non-viability of the water sector and external factors like macro-economic changes, political reform and institutional arrangements were put forward (Saleth and Dinar 2004) with most commentators putting forward external factors as the dominant one. The underlying principles are two pronged – to bring in financial efficiency and foster increased participation with a reduced role of the State.

In the 1980s and 1990s, USAID and World Bank-supported projects initiated policy changes in the irrigation sector in Sri Lanka, recommending farmer participation and introduced charging fees, the first move towards commodification of agricultural water that did not succeed due to poor participation (Bandaragoda 2005). The 1990s saw increasing consensus on the need for a new comprehensive water resource management policy in Sri Lanka due to competition among water users and the multiplicity of institutions dealing with water resources. The policy development processes that were proposed lacked the Sri Lankan context and awareness about politics, media and the public. Since the inception of the policy development process in 2001, the formulation of the policy was the responsibility of the Water Resources Secretariat, funded by the Asian Development Bank and the focus was more often technical while key socio-economic issues failed to receive strategic attention (Ariyabandu in this volume). The donor influence in the formulation of Flood Action Plan ignoring local interest and expertise in Bangladesh is another clear case of external influence on formulating national policies in South Asia.

The introduction of the National Water Policy (NWP) 1987 and its subsequent revision in 2002 signified major shifts in guiding principles in the arena of water reforms in India. The key differences between the two policy documents are that the 2002 policy focuses on the development of an improved institutional framework that reflected the nature of reforms being envisaged and undertaken in states across the country (Cullet 2006). While allocating priority to different uses of water, the NWP 2002 emphasizes the physical and financial sustainability of existing facilities by ensuring cost recovery of at least the operation and maintenance charges for providing the service initially and a part of the capital costs subsequently (Sampat 2007). Following this, many states also introduced state water policies, emphasizing private sector participation, tariff revisions, full cost recovery and creation of independent regulatory authorities.[6] In terms of governance changes, NWP 2002 (GOI 2002b) stated that:

> Management of the water resources for diverse uses should incorporate a participatory approach: by involving not only the various governmental agencies but also the users and other stakeholders, in an effective and decisive manner, in various aspects of planning, design, development and management of the water resources schemes. Necessary legal and institutional changes should be made at various levels for the purpose, duly ensuring appropriate role for women. Water Users' Associations (WUA) and the local bodies such as municipalities and gram panchayats (GPs) should particularly be involved in the operation. Maintenance and management of water

Infrastructures/facilities at appropriate levels progressively, with a view to eventually transfer the management of such facilities to the user groups/local bodies (GOI 2002b).

This typifies the sector reforms as advocated by the multi-lateral donors where the paradigm shift is reflected in government documents like the Eighth Five-Year Plan in India (1992–97) that emerged at the time of neoliberal reform. This introduced the concept of water as a commodity that should be allocated based on effective demand, cost recovery and managed by local organizations. The implementation of sectoral reforms includes suggestions for unbundling (separation of source, transmission and distribution), creation of independent regulatory institutions to free the sector from political intervention, increasing tariffs, cost recovery for financial sustainability, elimination of subsidies, retrenchment, privatization and Public-Private Partnerships (PPPs) and allocation of water, based on market principles (Dharmadhikary and Dwivedi 2009). It was also noted that sector reforms, while forming a significant aspect of the paradigm shift in the role of the state were effected relatively quietly and systematically with no legislative provision to effect this redefinition of the role of the state (Sampat 2007). The notion that water is an economic good is another major policy shift where all water services must be based on the principle of cost-recovery, particularly in a situation where the provision of drinking water as well as irrigation water is substantially subsidized (Cullet 2006). Three suggestions presented to address the issue of subsidies are – (*a*) divesting state from responsibilities other than policy making; (*b*) engaging private sector for provision of services, and (*c*) establishing independent regulatory agencies (Wagle and Warghade 2009).

The paradigm shift in large irrigation systems by irrigation law reforms is the idea of Participatory Irrigation Management (PIM)[7] whereby the responsibility of running the tertiary systems rests with local people's institutions like the Water Users Associations (WUAs) instead of the previous practice where the government does all the work such as construction, maintenance and allocation of water to farmers. Irrigation laws in the country vest the control and ownership over water resources in the state which addressed the issues of equity and rights, regulation and conservation, governance, management and finance, and water pricing. The WUA has now gained primacy within the institutional and legal reforms framework within the country (Cullet and Madhav 2009). In the Indian context, Cullet (2006) suggested that the shift entails a balance of benefits and burdens to WUAs as they benefit from (*a*) more assured water supply (since it is the authority's duty to supply the amount of water they are entitled to receive), (*b*) more control over water allocated to them and (*c*) the right to use groundwater in their command area on top of the entitlement they receive from canals. The burden comes from (*a*) responsibilities like regulation and monitoring of water distribution among WUA members, (*b*) assessment of members' water shares, (*c*) the responsibility to maintain an equitable supply of water to members, (*d*) the collection of service charges and water charges, (*e*) the carrying out of maintenance and repairs to the canal system and dispute resolution among members. Although WUAs are not commercial entities, they have to be economically viable and financially independent. Water Users Associations are also meant to be governed and controlled by people who pay for the services of the association (Cullet 2006). It is not clear how the policy addresses the concerns of those who might not be in a position to pay. A study from Rajasthan pointed to the functioning of WUAs and the challenges they have faced in the operation of irrigation systems. This raises serious questions about their ability to fulfil the ambitious responsibilities entrusted to them. This situation is compounded due to the lack of commitment from the state government to genuinely decentralize water governance to local communities. The issue of non-inclusion in irrigation institutions commonly discussed in the PIM debates in large irrigation systems was reflected even in small structures of water storage as well. The study concludes that shift in governance is in fact 'managerial' in order to bring in operational efficiency and is not oriented towards 'democratization', to make 'beneficiaries' and not to make rights-bearing 'citizens' (Narayanan and Kamath 2012).

In India, the provision of drinking water is primarily the responsibility of the states. With regard to policies followed by the Government of India, the first major sign of the desire to introduce reforms can be found in the revision of Accelerated Rural Water Supply Programme Guidelines in 1999–2000 which marks the beginning of the Sector Reform Projects, introducing community based management of rural water supply that is demand-based where users get the service they want and are willing to pay for (James 2004). The paradigm shift reflected in the water and sanitation sector when the following principles were introduced: (*a*) adoption of demand responsive approaches based on empowerment, to ensure full participation in decision making, control and management by communities, (*b*) shifting the role of governments from direct service delivery to that of planning, policy formulation, monitoring and evaluation, and partial financial support, and (*c*) partial capital cost sharing, in either cash or kind or both, and 100 per cent responsibility of O & M by users (GoI 2002). At the national level it is the Rajiv Gandhi National Drinking Water Mission under the Ministry of Rural Development which has been the key institution with regard to the development of policies and administration of the rural drinking water sector (Panicker 2007).

About the legal aspect, since water is a state subject, each state has relative freedom to evolve its own policy and legislative frameworks with due regard to its context. The existence of a number of laws in various states and the technical nature of law makes it difficult for practitioners, activists and other stakeholders to understand and use law for the benefit of the society and environment (Koonan Sujith 2010; Iyer 2009). In the ongoing debate about water as a right, a need or a commodity, the distinction between the individual's need for water and water as an economic good is debated. Adequate access to safe water must be treated as a right similar to other human rights enshrined in the Universal Declaration of Human Rights. The economic value of water could come to mean simply its partial value as a commodity that led to policy prescriptions such as the development of secured property right is *sine qua non* for efficient and sustainable management of water resources. The loss of access to all the rights in the bundle that are not reflected in the commodity market is a legitimate cause for concern. The Constitution of India does not explicitly address a citizen's right to water. Nevertheless, the Supreme Court has recognized right to water as part of the right to life generally, and has supported the public trust doctrine through Article 21 which assures life and personal liberty to all citizens (Iyer 2009; Koonan Sujith 2010). C. R. Bejoy and Ruana Rajapakse in this volume provide the legal nuances from a rights-based perspective in India and Sri Lanka, respectively.

An emphasis on cost recovery principles and a provision to allocate water on the basis of transferable water entitlements based on efficiency could be contrary to a rights-based allocation of water based on the principle of equity and economically weak users. For example, in the reform suggestions in Sri Lanka, the references to the need to "balance consumptive water uses such as irrigation, domestic and industrial, and non-consumptive uses such as hydropower, fisheries, recreation and aesthetics" led to a perception that the proposed measures were designed to favor private sector economic activities over traditional livelihoods, particularly agriculture (Rajapakse in this volume). She also observes that the policy thrust on "demand management," would affect Sri Lanka's rice farmers who were hitherto seen as important providers of food security to the nation and were encouraged to turn to less water-intensive crops for water to be allocated to higher valued uses over time. The imminent dangers became a reality in Maharashtra, India where the first Independent Water Resources Regulatory Authority (IRA) of Maharashtra Water Resources Regulatory Authority (MWRRA)started functioning in 2005. A study by PRAYAS conducted in 43 irrigation projects in the state clarified that about 2 lakh hectares of irrigation potential is lost due to diversion of water from agriculture to non-agricultural uses. While 50 percent of the diversions account for industrial use, the rest satisfy the increasing demand of domestic use in cities. The reallocations are done as per the state water policy that provides first and second priorities to domestic and industrial use, respectively. The reallocations are now being turned into 'tradeable entitlements', as per the water reform law (Warghede 2009).

A discussion on water sector reforms is not complete without a word on IWRM,[8] an approach that collects many normative concerns in water governance until now. While the concept internalizes environmental, social and human concerns, and enlarges the planning horizon from isolated projects to units such as a river basin, it is criticized as merely a refinement of the engineering tradition with an in-built tendency towards centralization and following a neoliberal economic logic that is inherently exclusive (Iyer 2007; Jairath 2008). In the industrialized countries, the concerns of environmental sustainability and economic efficiency are dominant whereas the plural South is more engaged in its industrializing mode, which involves the control of water resources to enhance productive uses of water? Beyond this water demand difference, the interpretation based on Dublin principles (especially "water as an economic good") has been consistently questioned from the South. As seen earlier, water allocation and associated rights to water is imminently political (Allan 2006) and the future water management scenarios would have to seriously look at questions of allocation which is a political process. The forthcoming discussions argue for a shift in favor of the neglected dimensions of social justice, equity and environmental sustainability which is required in place of the conventional arguments of hydraulic mission.

Implications of the Paradigm Shifts

Even before the reforms, the stratified social structure in South Asia had been exclusionary in terms of access to water. The documented history shows how technological innovations in drinking water supply begun elsewhere and under specific circumstances, initially made their way to serve a handful of elite families and their clientele in Kathmandu in the late 19th century, and over the last century trickled down to much of the peasantry (and secularized water), it is still excluding poor peasants and the bulk of the artisan classes and dalits (Sharma 2003; Sharma 2001). Studies in India revealed inequalities on the basis of class, caste, gender, ethnicity, etc as barriers to inclusion (Joy and Paranjape 2004; Narayanan and Kamath 2012; Shah 2003; Shah 2001). It is within this socio-economic and political structure that water 'development' interventions were done in South Asia. However, interventions like piped water were argued to be considered by people of becoming "modern" as well over reducing drudgery and hence treated as inevitable (Sharma in this volume).

About the democratic content of reforms, concerns were raised that citizens can no longer hold the Indian state directly accountable for securing basic services for all citizens. In the economic logic of this paradigm shift, ability to pay, in other words being a "user" seems to be the new minimal criteria for access to services, and thus, for the privileges of citizenship. In effect, this implies a privatization of service provision through divestment of state responsibility (Sampat 2007). An economic justification is that public provision and subsidized water charges belonged to the surplus era of water development which insulate water sector from the market forces. Water users, who were the beneficiaries, have now become 'customers' or 'clients' in the water scarcity era (Saleth 2004).

Legally, the main novelties introduced through recent legislations resulted in setting up of independent water regulatory authorities in India which are meant to take away part of the powers of government bodies and provide similar services without political interference, thus to de-politicize, and allocate water to various sectors (Cullet 2006) to be regulated and managed on a commercial basis like any other market good (Koonan Sujith 2010). The equity implications of these are emerging in various contexts in India.[9]

Through the introduction of shared cost and water tariffs in the schemes of sector reforms, only those with adequate resources are able to access water while others remain outside the purview of the scheme. Thus, those who cannot afford to pay the initial cost and contribute to the cost sharing are left to fetch water from already existing sources (Sampat 2007). This may lead to a significant and growing section of marginalized people being excluded from provision of water service; in other words increasing numbers of 'unreached' people, be it in rural or urban areas i.e. the continued prevalence of inequity in the distribution of water management of water resources has critical implications for people's lives and livelihoods for overall economic development and social prosperity (Golam and Chowdhary 2010).

In the case of large irrigation projects, we have earlier seen that extensive and possibly burdensome powers are extended to WUAs to not only manage the infrastructure but also to provide an institutional framework that equitably provides all the services that a public authority would provide. However, the question here is how an association of land holders that has no democratic legitimacy can ever perform all these tasks in an equitable and sustainable manner for its members and for the broader society around it (Cullet 2006) leading to the possible exclusion of the landless, women and lower castes (Narayanan and Kamath 2012). The ongoing reforms adopt a limited understanding of the terms Decentralization and Participation. For instance, Decentralization in the context of irrigation law reforms means formation of water users association. However, it does not mean giving more powers and resources to democratically elected/constituted bodies such as panchayat and gram sabha. In the same way, Participation refers to participation of land owners with the exclusion of landless people (Koonan Sujith 2010). The implications of the paradigm shifts in policy and governance is the major thrust and common thread throughout the contributions in this volume. The next section examines the responses to the water governance shifts from the sphere of civil society.

Civil Society Responses in Water Governance

In the current era of neoliberal dominance, the state, market and civil society are spheres that deem to exercise equal bearing in governance, undermining the State's dominant role historically. It would be interesting to examine the concept and practice as well as the changing role of civil society in water governance. A quick review of the (much debated) concept and contemporary issues related to civil society and placing NGOs within this fluid debate is attempted first.[10] The second subsection distils the observations on these issues from the contributions in this volume.

Concept and Disaggregation of Civil Society

The conceptual discussions on civil society are often marred with some degree of confusion and a lack of clarity. The Centre for Civil Society at the London School of Economics has adopted the following working definition for civil society which is not to be interpreted as a rigid statement,

> Civil society refers to the arena of un coerced collective action around shared interests, purposes and values. In theory, its institutional forms are distinct from those of the state, family and market, though in practice, the boundaries between state, civil society, family and market are often complex, blurred and negotiated. Civil society commonly embraces a diversity of spaces, actors and institutional forms, varying in their degree of formality, autonomy and power. Civil societies are often populated by organizations such as registered charities, development non-governmental organizations, community groups, women's organizations, faith-based organizations, professional associations, trade unions, self-help groups, social movements, business associations, coalitions and advocacy group.[11]

This inclusive definition in its broadest sense is the conceptual space outside family, market and state and can exhibit such diverse institutional forms. However, the tripartite division between the state, market and civil society is often problematic with the relative fluidity of boundaries between the spheres. Also, the treatment of civil society is often excessively normative – seen as a source of "good," distinct from a "bad" imputed to the state and market (Bebbington et al. 2008).

A Gramscian understanding of civil society constitutes an arena in which hegemonic ideas concerning the organization of economic and social life are both established and contested. In this formulation, the state and civil society are to be mutually constitutive rather than separate, autonomous entities, with both formed in relation to historical and structural forces. He was centrally concerned with explaining the failures of both liberalism and socialism, and the role that counter-hegemonic movements within civil society might play in promoting social and revolutionary change. The concept of civil society was suggested to be moribund when both the liberal and Marxist conceptions of social change and development models of state-led modernization disintegrated. The argument is that both the theorists disillusioned with ideas of socialism and also those who are critical of the retreat of welfare state through the neo-conservative reaction in the 1980s that led to atomization of society, argued for a pluralist tradition with the revival of associative initiatives of non-state organizations in civil society. In some other cases like in India, where stable democratic institutions are in place, authors expressed dissatisfaction with the social consequences of democracy and the failure of its party system and also demanded a civil society that needs to bring in accountability and transparency in governance (see the contributions in Tandon and Mohanti 2003).

Civil society and the place of NGOs within it must be understood at two levels – at the level of ideology and theory, the notion of civil society has flourished most within either the neoliberal school that advocates a reduced role of the state or a post-Marxist/ post-structural approach that emphasizes the transformative potential for social movements within civil society. At the conceptual level, civil society is usually treated in terms of associations (so called civil society organizations) or as an arena within which the ideas about the ordering of social life are debated and contested.[12] This understanding is the starting point of the functional role of this sphere, which is more of contestation, especially, challenging the dominant notions and embrace normative ideals difficult to practice within the other two spheres of state and market.

To explore the more difficult question of NGOs and their role in bringing an alternative perspective, especially to hegemonic ideas, Bebbington et al. (2008) identify three phases in history where the first one up to mid 1960s was without any such trend. The second phase up to mid 1980s witnessed NGOs contributing to political struggles around national independence and various socialisms as well as to intellectual debates on dependency, structuralist and broadly Marxian interpretations of the development process, which contributed to a notion of "alternative development." The process also contributed to a process of collaboration among movements, supportive institutions and NGOs for a struggle against hegemonic and repressive structures.[13] The bulk of the contestation revolved around political rather than economic structures. The third phase in the last two decades is marked by the creation of the World Trade Organisation (WTO), the neoliberalization of social democracy, the end of global communism, military enforcement of liberal democratic process, the joint project of liberal democracy and free trade seems to have become increasingly clear making it difficult for NGOs to think or act outside of this neoliberal box. The major trend noted in the neoliberal phase is the call for deepening democratic institutions by civil society without challenging its economic logic.

With the deepening of both democratization and the neoliberal agenda in developing countries with a shift of donor priority and state funding to poverty reduction agenda,[14] NGOs were reduced to "public service contractors", especially with the reduction on funding, oriented towards systemic alternatives and challenging hegemonic ideas. The specific forms of intervention have also involved the increased channeling of resources through NGOs with bundling of particular rules and ideas (related to neoliberalism and security) regarding how they must be governed and contribute to the governing of others. The emphasis has shifted to more instrumental goals like cost recovery, charging for services, professionalized staff relationships, the dominance of competition and the rise of tenders, which blurred civil and market logic and eroded NGOs' potential as agents of systemic social and political change.[15]

Even with this bleak picture, when the water sector in India is going through the various reform processes, the response of NGOs and social movements has been critical in deepening the trend as well as identifying alternative models and enlisting greater participation of people in the management of public affairs both at the broader policy and implementation levels. Another author has put forward five models of governance with participation – first is the imperative to roll back the state; the second involves the contestation of state projects, practices and discourses contained in the social movements arguing for a radical participatory democracy; the third relates to NGOs that seeks to implement state initiatives in this field as public service contractor; the fourth model brings together the state and community in partnership, and the fifth refers to decentralizing the state and making them responsible for the local management of water related services and water bodies. All of these reinforce the neoliberal idea of the declining role of state in the water sector. In the era of liberalization, privatization and globalization, the development NGOs seem to be filling in the space left by both the state and radical social movements since there is a gradual withdrawal of the state from some of its traditional functions, and NGOs (as well as corporate bodies, either through public–private partnerships or privatization of service delivery) are stepping into those spaces. While NGOs are increasingly getting into service delivery and implementation of projects, there are still instances where they have challenged the existing policies and programs. There are also social movements which have carved a contested space for themselves, outside the space conceptually and operationally provided by the state. It is in this background that we have to see the civil society strategies in water governance in South Asia. The next sub-section examines how this paradigm shift and the civil society responses are reflected in the contributions in this volume.

Key Observations from this Volume

The eleven chapters in this volume present a comprehensive narrative on water governance in South Asia. There is an exclusive conceptual contribution on the nature of state and civil society in South Asia in the introductory section while the second section explicitly examines the policy process and the influence exercised by NGOs. The major paradigm shift in policy in India with a thrust to watershed management is reviewed in order to distil the contribution of NGOs in policy formulation and implementation in the third chapter. A subsequent chapter in this section illustrates the strategic approach of a successful NGO in India, where the fundamental rules of the state were skilfully used to influence policy changes. Contrary to this, the last section titled as social movements portray initiatives that challenged the state. Their organizational forms were more fluid with the affected people and NGOs consolidating a coalition to challenge the policy and governance changes. They created a difficult and contested space in water governance external to that provided by the state. The fourth section provides the policy reversals and contestations to that by NGOs in Sri Lanka and Bangladesh. The third section specifically examines governance changes and policy shifts through donor influence in the provision of water and sanitation services in India, Nepal and Sri Lanka.

In Section 1, Mahfuzul H. Chowdhury's chapter is more of a conceptual contribution painting the larger picture of state–society relations in South Asia. It is different from most other chapters and the only one that presents a neoliberal view of governance with a mainstream understanding of the term following international financial institutions. The author argues that decentralized governance structures offer better space for civil society interactions compared to more centralized authoritarian states and thus present a key and common challenge for democratic governance in South Asian countries. The author's attempted definition of civil society follows the mainstream-neoliberal assumptions and views it as a sphere to enhance efficiency in governance. It similarly categorizes civil society according to their allegiances to the state and also challenges the democratic nature of CSOs and NGOs in South Asia.

In Section 2, the chapter by K. J. Joy analyzes the scope and constraints of civil society action in the context of watershed development policies and programmes in India. He classifies CSOs into those that work within the space provided by the state (and hence do not challenge existing social relations of production), and those that challenge them with a transformative agenda. The CSOs involved in watershed development are observed to be mainly those who want to work within the space provided by the state or within the limits imposed by the present social relations. Donor agencies also significantly impact watershed policies and programs since funding is coupled with certain conditions and guidelines. This also partially explains why civil society action could not fully utilize the space available within the system to promote normative goals like equity, sustainability etc. Vasudha Chhotray in the same section paints a more nuanced picture of a successful NGO in Madhya Pradesh, India. She illustrates the way NGOs view the state and negotiate a space to undertake socially transformative development. The insights from this one microcosm points to the character of Indian civil society and the opportunities and constraints available to NGOs, given their often unique position in the interface between governments at different levels (both elected representatives and bureaucrats), local communities and foreign donors.

The third section has three contributions from Nepal, India and Sri Lanka on the influence of donors on the water and sanitation services in South Asia, which is linked to the agenda of "millenium development goals". With "safe water" replacing "piped water" as the raison d'être for intervention, Sudhindra Sharma's contribution examines the history of water supply and sanitation in Nepal. He argues how the idea of piped water modernity, which was the prerogative of a few elites in Kathmandu has been translated into the imagination of Nepali villages since it was not only for reducing the drudgery, but for becoming "modern" too. The other two chapters in the section illustrate the consequences of such a process, which are increased foreign debt, technology dependence and institutional weakening. The Sri Lankan case study is primarily a whistle-blower to the proposed large-scale water supply and sanitation project to be implemented by the Kandy municipality in Sri Lanka, funded by the Japan Bank for International Cooperation. The civil society responses related to the contributions in Section Three are also varied. In Nepal, there is no mention of any questioning or reflection while in the Kerala case, civil society is practically absent in the urban sector. Similarly in Sri Lanka, a social movement that emerged in the beginning got fizzled out in due course of time. The Sri Lanka and Kerala cases brought out the bloated expenditure and high tariff rates to recover the operation and maintenance costs and possible future burden, particularly to lower and middle income households due to the high cost of donor supported projects. The technological appropriateness, economic viability, social acceptance, political feasibility and thus the sustainability of the projects are never transparently discussed in any forum. The only presence of NGOs was that as "support organisations" in rural water projects, which is basically of a public service contractor during the construction phase. Here the contestation at the level of ideas is not possible and in fact NGOs become local careers of ideas of cost recovery and "demand responsive approach". Hence the role of NGOs as visualized in Section Three is problematic and contrary to the assigned or perceived role of civil society.

The fourth section involves the policy process and contestations by NGO coalitions to specific changes in policy. The policy changes facilitated by the Asian Development Bank (ADB) and the contestations in Sri Lanka are included in two contributions and the third one discusses the protests against the Flood Action Plan in Bangladesh. Rajindra Ariyabandu raises some fundamental issues on the role of civil society in the policy process in Sri Lanka, more from the perspective of the state. While acknowledging that the policy process lacked transparency and accountability, he argues that NGOs representing the civil society misrepresented some of the facts in their agitation against the water policy process. Due to inadequate information and misunderstanding, policy attracted severe criticisms since many of the officials who were governing the reform process were ill-equipped to face the criticism. He identifies three types of civil society groups in the protests − (*a*) those concerned about the environmental implications and threats to livelihoods, (*b*) those who wanted to expose the nexus between global and local water traders, and (*c*) those who wanted to derive political mileage from the debate. While these three groups used the media to their advantage, the media itself used the issue to their own advantage in isolating and attacking the political leadership at different times of the policy process. According to him, the civil society and media campaign against the reform process was expected to identify gaps and challenges in the water policy debate, inform the public and work towards sustainable water resource management. However, their actions have left water resources more vulnerable to exploitation. Nevertheless, a civil society perspective would present the protest movement as a success that prevented an externally facilitated neoliberal policy agenda being adopted in the country. Instead, Ruana Rajepakse presents a case of how a pro-public judiciary can stand to protect the fundamental rights of citizens in the cases of rights to water, rights to livelihood and rights to protect water sources from pollution. The author also highlights the role of the CSOs in protecting the common law principles and human rights to water during the water policy development process supported by multi-lateral donor agencies in Sri Lanka. Hamidul Haq's chapter presents the NGO (coalition) strategies of lobbying, networking and indirect advocacy in the actions against the Flood Action Plan (FAP) in

Bangladesh where Bangladeshi NGOs worked for greater access to policy information and dialogue with the policy/decision makers. They built networks with national and international media, environmental activist groups, foreign governments, research organizations, lobby groups, academic institutes and international NGO donors. They made the FAP information available to the local communities and emphasized the value of local expertise and experience in traditionally dealing with floods. The Bangladeshi NGOs could effectively influence the European media to raise a critical mass of dissent against the FAP and to urge member states to reconsider their support to the Plan.

The last section in the volume presents the case of two social movements which proved that there is enough space within the legal structure to challenge and how it was used by the most marginalized actors to assert their rights on water with a rainbow coalition of support from various strata of society, even without the help of major political parties. While NGO influence is through indirect lobbying, social movements in turn use more explicit and overt strategies of political mobilizations and demonstrations to put forward their demands. The chapter by C. R. Bijoy argues that the Constitution provides sufficient scope for the emergence of a democratic participatory governance system at its fundamental level – the community – especially in the case of groundwater and minor water bodies, the fundamental base on which the structure of water governance could be built upon. This is evidenced by how the adivasis and Dalits of Plachimada in Kerala who are the poorest, mostly landless and illiterate agricultural workers, remained at the forefront of the struggle against one of the biggest corporate entities like Coca Cola. They received spontaneous support from diverse sections, from the extreme right to the extreme left, from neo-Gandhians to environmentalists, and from children to peasants across the state. In Chapter 10, Suhas Paranjape and Seema Kulkarni discuss the case of a south Maharashtra water movement that raised very difficult demands such as equitable distribution of water within a very unequal social structure. The movement gained support of the masses by the transparent strategies and uncompromising stance of the leadership. The combination of factors, including experts' help to ensure feasible alternatives, innovative ways of making demands, organizational strategies generating a mass base (including the landless, small peasants, women and the working class) has resulted in numerous concrete and long-term gains in expanding and defending water rights, especially for the poor.

A clear trend emerging out of the case studies is the difference between the strategies of NGOs and social movements. NGOs work within the space provided by the state and organizations range from service providers on behalf of the state (or outside it) and others who skilfully negotiate a policy space by working with the state. In contrast, social movements carve a contested space to enlarge the bundle of rights. It is a process of adversarial politics by, as well as on behalf of, the otherwise powerless, deprived and disadvantaged sections. Much more than an NGO, social movements have an expressive energy and militancy that sustains them. It is also clarified that it is difficult to imagine an NGO achieving the same objectives, since they would typically avoid confrontation and radical street action. An NGO's milder tactics, aimed at achieving specific limited aims, would be unlikely to make sustainable gains of their action, especially with a change in state/political power.

Notes

1. The manuscript for this volume was finalized in 2011. The work of the working groups of the Planning Commission was only initiated at that time. Many of the water sector policy proposals in the Twelfth Plan document add up to a paradigm shift in thinking. This chapter has not included these discussions as an afterthought while finalizing the manuscript.
2. See the excellent contribution by Tushaar Shah outlining the historical development of irrigation systems in South Asia, which suggests that in the past 40 years the South Asian countries witnessed more development in irrigation than preceding two centuries and forms a major bulk (90 out of 300 million ha) of area under irrigation in the world (Shah 2009).
2. For a passionate account of justification of the need for large water infrastructure for India see Briscoe and Malik 2006; for Pakistan see Briscoe and Qamar 2006 and for Bangladesh see Briscoe 2001.
3. Schedule 7, List 2, Entries 17, 21 of the Constitution of India.

4. Schedule 7, List 1, Entries 24, 27 and Article 262 of the Constitution of India.
5. See Byres 1998, for the Indian case to neo-liberal shift in policies and the changing role of the state. Toye (1993) detailed the coming back of old neo-classical ideas in the wake of this crisis, which is then termed as a "counter-revolution" in development economics.
6. Karnataka brought out its water policy in 2002 followed by Madhya Pradesh and Maharashtra in 2003, Odisha (2007), Andhra Pradesh (2009) and Uttar Pradesh (revised the 1999 policy in 2009). Regarding the creation of IRAs, Maharashtra was the first to create the MWRRA in 2005 followed by Arunachal Pradesh (2006), Uttar Pradesh (2008) and Andhra Pradesh (2009).
7. In India, a number of states like Andhra Pradesh, Madhya Pradesh, Odisha and Rajasthan that introduced WUA legislation show similarity in basic principles although they have been adopted at different points in time and the schemes proposed have evolved over time even though the basic principles are fairly similar in each situation (Cullet 2006). While the act provides a decentralisation scheme towards farmer involvement in irrigation at the local level, it also gives significant powers to the Independent Water Resources Regulatory Authority (IRAs) or other designated authorities (Cullet 2006, Sangameswaran 2006)
8. A popular definition refers IWRM as a process which promotes the coordinated development and management of water, land and related resources, in order to maximize the resultant economic and social welfare in an equitable manner without compromising the sustainability of vital ecosystems (GWP 2000: 22). The definition clarifies its normative content with the collection of concerns bundled within the concept. Environmental fundamentals such as the hydrological logic of the river basin and economic fundamentals relating to the value of water are central to the implementation of IWRM. For a comprehensive treatment of the concept and its practice in South Asia, refer Mollinga et al., (ed) 2006.
9. For the related issues of the Maharashtra experiences of reforms (specifically with the introduction of Independent Regulatory Authorities) refer to Warghede 2009.
10. A comprehensive theoretical treatment of the concept in a historical sense and nuances of the debate globally has been compiled by Kaviraj and Khilani. For an Indian perspective, see the contributions in Tandon and Mohanti (2009). An excellent recent compilation, especially the question of NGOs as civil society, especially the constraints and scope of NGOs as development alternatives in the era of neoliberalism is available at Bebbington et al. (2008).
11. http://www.lse.ac.uk/collections/CCS (accessed November 12, 2013).

12. Edwards, 2004 quoted in Bebbington et al., 2008.
13. The noted examples are as in the case of Philippines, South Africa and El Salvador. European co-financing resources were given without many questions being asked, in order to channel resources to oppositional movements via NGOs without explicit, traceable government knowledge, whereas conservative forces – notably USA – supported the hegemonic forces and ideas against which these NGOs and political movements were struggling (Hulme 2008).
14. Kees Biekart (2008) shows that donor funding has concentrated to a smaller set of countries with European agencies moving away from areas such as rural development, agriculture and the environment with increased attention to topics such as migration, conflict resolution, peace-building and trade issues.
15. Fowler (2005) quoted in Bebbington et. al., 2008.

References

Allan, J. A. 2006. "IWRM: The New Sanctioned Discourse?," in Peter P. Mollinga, Ajaya Dixit and Kusum Athukorala, *Integrated Water Resources Management: Global Theory, Emerging Practice and Local Needs*. New Delhi: Sage.

Apte, Tejaswini. 2001. *Water Rights Land Reforms and Community Participation: The Pani Panchayat Model for Sustainable Water Management*. Study done for the Lokayan-Kalpavriksha on Conflicts over Natural Resources and their Resolution, Mimeo.

Atwood, Donald. 2007. "Small is Deadly, Big is Wasteful: The Impact of Large Scale Industrial Systems in Western India," in Amita Baviskar (ed.), *Waterscapes: The Cultural Politics of a Natural Resource*, pp. 11–39. Uttarakhand: Permanent Black.

Bandaragoda, J. D. 2005. "Relevance of River Basin Management as a Strategy for Sri Lanka Water Sector", Consultation of River Basin Management. Colombo: Lanka Jalani.

Bandyopadhyay, J. 2002. "A Critical Look at the Report of the World Commission on Dams in the context of the debate on large dams on the Himalayan Rivers," *Water Resources Development*, 18: 127–45.

Bebbington Anthony, J., Samuel Hickey and Diana C. Miltin. 2008. "Introduction: Can NGOs Make a Difference? The Challenge of Development Alternatives," in Anthony J. Bebbington, Samuel Hickey and Diana C. Miltin (eds), *Can NGOs Make a Difference? The Challenge of Development Alternatives*, pp. 3–37. London & New York: Zed Books.

Briscoe, J. 2001. "Two Decades of Change in a Bangladeshi Village," *Economic and Political Weekly*, 36(40): 1–8.

Briscoe, J. and Malik, R. P. S. 2006. *India's Water Economy: Facing a Turbulent Future*. New Delhi: Oxford University Press.

Briscoe, J. and Qamar, U. 2006b. *Pakistan's Water Economy: Running dry*. Karachi: Oxford University Press.

Byres, T. J. 1998. "Introduction: Development Planning and the Interventionist State Versus Liberalisation and the Neo-Liberal State: India, 1989–1996," in T. J. Byres (ed.), *The State, Development Planning and Liberalisation in India*, pp. 1–35. New Delhi: Oxford University Press.

Cernea, M. M., (ed.). 1999. *The Economics of Involuntary Resettlement: Questions and Challenges*. Washington D.C.: The World Bank.

Chowdhury, J. U., M. Rahman and M. Salehin. 1997. *Flood Control in a Floodplain Country: Experiences of Bangladesh*. Islamic Educational, Scientific and Cultural Organization (ISESCO), Rabat, Morocco.

Cullet Philippe and Roopa Madhav. 2009. "Water Law Reforms in India: Trends and Prospects," in Ramaswamy R. Iyer (ed.), *Water and the Laws in India*. New Delhi: Sage.

Cullet, Philippe. 2006. *Water Law Reforms: Analysis of recent development:* New Delhi: IELRC.

D'Souza, Rohan. 2008. "Framing India's Hydraulic Crisis: The Politics of the Modern Large Dam," *Monthly Review*, http://www.monthlyreview.org/ 080811dsouza.php (accessed on March 3, 2011).

Dharmadhikary, Shripad and Gaurav Dwivedi. 2009. "Water Sector Reforms and their Implications in Madhya Pradesh". Paper presented in the International Conference on Water Resources Policy in South Asia, December 17–20, 2008. Colombo, Sri Lanka. Organized by South Asia Consortium for Interdisciplinary Water Resources Studies (SaciWATERs) Mimeo.

Government of India (GOI). 2002a. *Water Supply and Sanitation: A WHO-UNICEF Sponsored Study,* Planning Commission, http://planningcommission.nic.in/reports/genrep/wtrsani.pdf (accessed on March 21, 2011).

———. 2002b. National Water Policy. Ministry of Water Resources, Government of India, http://mowr.gov.in/writereaddata/linkimages/nwp20025617515534.pdf (accessed on March 20, 2011).

———. 2006. "*From Hariyali to Neeranchal: Report of the Technical Committee on Watershed Programmes in India*," Department of Land Resources, Ministry of Rural Development. New Delhi: Government of India.

Golam, Rasul and A.K.M. Chowdhury. 2010. "Equity and Social Justice in Water Resources Management in Bangladesh", Gate Series 146. London: International Institute For Environment and Development

Global Water Partnership (GWP). 2000. Towards Water Security: A Framework for Action. Stockholm: GWP.

Hossain, Monowar. 1994. "Existing Embankments", in K. Haggart (ed.) *Rivers of Life*, Dhaka: BCAS.

Hulme, David. 2008. "Reflections on NGOs and Development: The Elephant, the Dinosaur, Several Tigers but No Owl," in Anthony J. Bebbington, Samuel Hickey, Diana C. Miltin, *Can NGOs Make a Difference? The Challenge of development Alternatives*, pp. 337–45. London & New York: Zed Books.

Iyer, Ramaswamy. 2007. *Towards Water Wisdom: Limits, Justice, Harmony*. New Delhi: Sage.

———. 2009. *Water and the Laws in India*. New Delhi: Sage.

Jairath, Jasveen. 2008. "A Plea for Unpopular Wisdom," *Economic and Political Weekly*, 43(7): 32–35.

Joy, K. J. and Suhas Paranjape. 2004. "Technical Report: Watershed Development Review: Issues and Prospects." Bangalore: Centre for Interdisciplinary Studies in Environment and Development (CISED).

Kees, Biekart. 2008. "Learning from Latin America: Recent Trends in European NGO Policy Making," in Anthony J. Bebbington, Samuel Hickey, Diana C. Miltin (eds), *Can NGOs Make a Difference? The Challenge of Development Alternatives*, pp. 71–89. London & New York: Zed Books.

Kerr, John. 2002. "Watershed Development Environmental Services, and Poverty Alleviation in India," *World Development*, 30(8): 1387–1400.

Koonan Sujith. 2010. "Water Law and Policy in India: Reforms and Capacity Building", New Delhi, Environmental Law Research Society, Mimeo.

Mollinga, Peter P., Ajay Dixit and Kusum Athukorala. 2006. *Integrated Water Resources Management: Global Theory, Emerging Practice, and Local Needs*. New Delhi: Sage.

Narayanan, N. C. and Lalitha Kamath. 2012. "Rural Water Access: Governance and Contestation in a Semi-Arid Watershed in Udaipur, India," *Economic and Political Weekly*, 48(4), 65–72.

Panicker, Meena. 2007. "State responsibility in the Drinking Water Sector: An Overview of the Indian Scenario". New Delhi: International Environmental lan Research Centre.

Patankar, Bharat and Anant Phadke. 2008. "Dams and displacement: A Review," in K. J. Joy, B. Gujja, S. Paranjape, V. Goud and S. Vispute (eds), *Water conflicts in India. A million revolts in the making*. New Delhi: Routledge.

Patil, R. B. 2000. *Pani Panchayat in Pune district*. New Delhi: Rawat Publications.

Phadke, A. and B. Patankar. 2006. "Asserting the Rights of the Toiling Peasantry for Water Use," in, P. Mollinga, A. Dixit and K. Athukorala (eds), *Integrated Water Resource Management: Global Theory, Emerging Practice, and Local Needs*. Water in South Asia Series. New Delhi: Sage.

Saleth, M. R. and A. Dinar. 2004. *The Institutional Economics of Water: A Cross-country Analysis of Institutions and Performances.* Cheltenham, UK: Edward Elgar.
Sampat, Priti. 2007. "'Swa'-jal-dhara or 'Pay'-jal-dhara – Sector Reform and the Righ to Drinking Water in Rajasthan and Maharashtra' Law," *Environment and Development Journal*, http://www.lead-journal.org/content/ 07101.pdf (accessed on February 10, 2011).
Sangameshwaran, Priya, EPW: May 27, 2006. "Equity in Watershed Development – A Case Study in Western Maharashtra," *Economic and Political Weekly*, 41(21): 2157–65.
Sangameswaran, Priya. 2010. "Rural Drinking Water Reforms in Maharashtra – The Role of Neoliberalism," *Economic and Political Weekly*, 45(4): 62–69.
———. 2009. "Water Rights for the Landless in Western India: From Pani Panchayat to Water Entitlements," *European Journal of Development Research* 21: 195–212, http://www.palgrave-journals.com/ejdr/journal/v21/n2/full/ejdr2008119a.html (accessed on November 10, 2013).
Shah, Amita. 2001. *"Who Benefits from Participatory Watershed Development"*, Gate series 97. London: International Institute For Environment and Development.
Shah, Esha and Praveen Singh. 2007. "Community-Based Natural Resource Management in Gopalpura, Rajasthan," in Ajit Menon, Praveen Singh, Esha Shah, Sharachachandra Lele, Shah Esha (eds), *Social Designs: Tank Irrigation Technology and Agrarian Transformation in Karnataka, South India.* New Delhi: Orient Blackswan.
Shah, Mihir, Debashis Banerji, Vijay Shankar and Prematesh Ambasta. 1998. *India's Drylands: Tribal Societies and Development through Environmental Regeneration.* New Delhi: Oxford University Press.
Shah, Tushaar. 2009. *Taming the Anarchy: Ground Water Governance in South Asia.* Washington: RFF Press.
Sharma, S. 2003. "Water in Hinduism: Continuities and Disjunctures Between Scriptural Canons and Local Traditions in Nepal," *Water Nepal: Water, Human Rights and Governance*, 9(1): 215–47.
———. 2001. *Procuring Water: Foreign Aid and Rural Water Supply in Nepal.* Kathmandu: Nepal Water Conservation Foundation.
Tandon, Rajesh and Ranjita Mohanti (eds). 2003. *Does Civil Society Matter?: Governance in Contemporary India.* New Delhi: Sage.
Toye, J. 1993. *Dilemmas of Development: Reflections on the Counter Revolution in Economics.* Oxford: Blackwell.
Wagle, Subodh and Sachin Warghade. 2009. *Independent Regulatory Agencies and Related Reforms Water Sector in South Asia: Implications for Social Policy.* Pune: Prayas.

Warghade, Sachin. 2009. "Assessment of Reform Policy Instruments for their Contribution to Empowerment and Equity in the Water Sector". Paper presented in Fourth South Asia Water Research Conference, Kathmandu, Nepal.

Watts, M. 1994. "Development II: The Privatization of Everything?," *Progress in Human Geography*, 18(3): 371–84.

World Bank. 1998. *Recent Experience with Involuntary Resettlement: Overview.* Washington D.C.: Operations Evaluations Department.

1
The State, Civil Society and Challenges to Democratic Governance in South Asia

Mahfuzul H. Chowdhury

Introduction

Governance is a rapidly growing subject and has been pursued in a vast area of academic and empirical interests. The concept of governance is becoming popular among scholars, as well as practitioners in the fields of democracy, development, international relations, globalization and the environment, yet there is no clear consensus on its meaning and definition. However, the concept has been applied in political science, public administration, public management, international organizations, international relations, political economy, corporate administration and environmental administration and studies. It is interesting to note how governance is gaining ground in making democracy, state, regional and international organizations effective in their dealings with the people and stakeholders, in producing and distributing public goods, organizing social and public interests, and managing common affairs in the society. Academically, also, it is fascinating to investigate how governance has been evolving, as self-governing networks of rules transcend national territories and even by-pass or undermine traditional government networks. So "the popularity of governance derives from its capacity–unlike that of the narrower term 'government' – to cover the whole range of institutions and relationships involved in the *process* of governing".[1]

Governance studies focus on the nature of state, society, and political institutions and their relations with civil society. They venture to identify the networks of rules which govern public affairs and promote common goals of a society. The focus on governance studies in political science and public administration is particularly on, among other things, democratization, accountability, transparency, decentralization, participation, and operational principles governing activities. Beyond academia, the concept has gained ground both in the operational activities of national and international organizations such as the World Bank, United States Agency for International Development (USAID), European Union and many others, including nation states, the corporate world, and civil society organizations.

This study intends to examine the nature of states, political structures, actors and institutions in South Asian countries and the space available in these nations for civil society, from the perspective of governance. More specifically, it will explore the existence of space available for non-governmental organizations (NGOs) and civil society to effectively engage with the state in areas of common public interest, as well as to see how to include people in all social and communitarian affairs for common benefit.

Literature on Governance

Although studies on governance are relatively new, innumerable articles have been published in many scholarly journals, and even specialized scholarly journals are now available on governance. One study found that between 1986-98, the Social Sciences Index recorded 1,774 articles published in several journals, but only in three years, between 1999-2004, the Index recorded 1,855 articles on the subject.[2] This increase in the number of publications on governance indicates the interest and importance that the subject has recently generated. Some donors, like the World Bank and USAID, have added qualifiers such as "good" to the concept of governance. Thus, "'good governance,' a term that came into vogue in the 1990s, with the World Bank leading the charge, has assumed the status of a mantra for donor agencies as well as donor countries for coordinating aid, depending upon the performance of the recipient government".[3] In fact, nowadays the concept of "good governance" is popular worldwide. According to A.W. Rhodes, "Governance refers to self-organising, inter-organisational networks characterised by independence, resource exchange, rules of the game, and significant autonomy from the state".[4] This approach to an understanding of governance is relevant for public administration, public policymaking and an implementation mechanism, and is important to make the activities of the government effective and efficient. But this approach has its limitations, as it is not clear how the self-governing rules would be developed, promoted and implemented, without specific governmental machinery, agencies or organisations. Hyden perceives governance as "the stewardship of formal and informal political rules of the game. Governance refers to those measures that involve setting the rules for the exercise of power and settling conflicts over such rules".[5] Anne Mette Kjaer applies this approach to comparative politics and political development. With this definition, the question of the role of government or governing agencies has also become an omnipotent issue, as without them, who and how the rules of the game would be developed and monitored remain uncertain.

Rosenau applies a similar concept of governance to international matters. He writes, "Global governance is conceived to include systems of rule at all levels of human activity – from the family to the international organization – in which the pursuit of goals, through the exercise of control has transnational repercussions".[6] Here also, the question of self-governing rules and regulations come to the fore regarding governance in the international context, but the issue of international authority or institutions to make the rules and monitor their enforcement remains unresolved. Notwithstanding this, the concept of governance or good governance has become synonymous with democracy, decentralization, transparency, accountability, people's participation, rule of law, networking of agencies and legitimacy of authority, and clearly it means moving away from the traditional concept of government.

The World Bank has identified six dimensions of good governance: "voice and accountability, political stability, government effectiveness, regulatory quality, rule of law, and control of corruption".[7] However, there are serious questions about the practice of these governance principles within the World Bank itself or by the Bank. The USAID and its Center for Democracy and Governance focused on "rule of law, elections and political processes, civil society and governance' for their activities and attached these factors to 'democratic principles such as transparency, pluralism, citizen's involvement in decision-making, representation and accountability," and concentrated its activities in five areas: democratic decentralization, legislative strengthening, government integrity, policy implementation and civil-military relations.[8] However, the concept of governance is still evolving, and there is no clear consensus on it among scholars and practitioners.

Governance in South Asia

In South Asian countries, the concept of governance or good governance is predominantly linked to the government. The basic perception is that governance is the function of government, and good governance means efficient functioning of government institutions. Popularly, governance is often linked to democracy, accountability, transparency, decentralization, containing corruption, and efficiency of government institutions, but within the conventional concept of government.[9] The distinction between the concepts of government and governance is not clearly understood, but governance is understood to be more wide-ranging than the government. The focus of governance discourse in South Asia is mostly concentrated around traditional institutions of the state, government and their functions, but not clearly on developing networks of self-governing rules which are autonomous from the government and state. It is obvious that for effective governance, a well-developed civil society is important to develop and enforce the self-governing networks of rules in the society and to govern civil society organizations. It may be worthwhile, therefore, to examine the nature of states and the scope of civil society activities in South Asia from the governance perspective.

Nature of State, Government and Political Actors in South Asia

South Asian states are varied in their nature, structural composition and political fabric, and therefore very interesting to investigate.

India, among the largest states in South Asia, is also the largest democracy in the world. It has maintained a democratic system, in spite of experiencing several crises since its independence from British rule in 1947. These include the humanitarian crisis of settling the refugees who came to India from Pakistan and people who left India for Pakistan due to the Partition of India, following widespread violence and riots; India's wars with its neighbors China and Pakistan, and the assassination of Mahatma Gandhi, the father of the Indian nation. None of these crises derailed Indian democracy, except for a brief period of Emergency, imposed in 1975–77 by the then Prime Minister Indira Gandhi.[10] India's continuing democracy has elevated it to a model among developing nations, negating the theory that countries or societies without a certain level of economic development cannot be democratic.

Nonetheless, the questions one needs to investigate on Indian democracy are: what is the nature of Indian democracy, how it manages its development needs, what is the urban-rural gap in socio-economic development, what is the state of social justice, equity, and deprivation of the socially backward communities or subaltern groups? All these are considered in the context of its recent economic growth, especially in the knowledge-based industries. Students and scholars of Indian democracy and South Asian politics also confront the question of how it is possible that India remains democratic, while its neighbor Pakistan, for most of its history, has been going through military rule in one form or another, and Bangladesh has experienced military rule for a significant part of its history and its democracy is frequently challenged. Sri Lanka, despite experiencing violent civil war for decades, has maintained a democratic political system since its independence, with certain deviations, but never experienced military rule until now. So, "unusually, India and Sri Lanka have maintained and consolidated a democratic system of government, though neither achieved this feat with an unblemished record".[11]

Nepal, traditionally a monarchy, has for some time had constitutional monarchy with the monarch exerting a strong influence on the political system. It became a republic country only in 2008, when the Maoists and civil society put up a fierce resistance to the absolute rule of King Gyanendra; the Maoists won the elections and abolished the monarchy. Interestingly, Bhutan is still transiting from monarchy to democracy, a process voluntarily initiated by the reigning King himself. The Maldives, a small island nation, has been under authoritarian rule and shows no sign of democratic transition.

Here we will focus mainly on India, Pakistan and Bangladesh, for reasons of size (territory, population and resource base), political dynamics and social diversity. India, relatively larger in population, territory and resources than Pakistan and Bangladesh, is considered the role model for others in South Asia with respect to governance. This is because of a range of factors, such as the federal or decentralized nature of the Indian state,[12] democratic political system, pluralist society and culture, network of nationwide local government, and a strong middle class. Obviously, India is at an advantage to adopt, execute and implement the governance principles and therefore an interesting case to study. Concurrently, however, it is also essential to explore the failures of Indian democracy which inhibit the practice of governance/good governance in the country.

India became independent in 1947 and a republic in 1950, but it also adopted certain policies which threatened democratic, liberal and pluralist principles. Most notable of these were a centrally planned, largely state-controlled economy, especially in industry, and as a corollary, it fostered the growth of a highly centralized, powerful bureaucracy that has been fairly inefficient and corrupt. This hindered the economic and social development of Indian society and led to the growth of a license *raj* in India. The gaps between urban and rural areas, between religious communities, and between the genders are still pervasive in Indian society.[13] The growth of poverty and economic stagnation were the inevitable results of Pandit Nehru's socialist economic policies.[14]

However, by the early 1990s, liberal economic reforms were initiated by his successors in the Congress leadership. Today, its near double-digit economic growth rate has left behind its "Hindu rate of growth".[15] In the 2006 ratings of global economic competitiveness, India ranked 43, Pakistan was 91 and Bangladesh was 99. Therefore, there is a consensus among scholars that political democracy has contributed very positively to the stability and continuity of the Indian political system. The number of people below the poverty line reduced to around 27 percent from 40 percent of the total population,[16] although the gap between urban-rural people has widened in the last 50 years.[17] Yet, there are regional variations of economic, social and democratic conditions in the country.[18] There are differences in growth rates and socio-economic indices among the various Indian states – Kerala, Maharashtra, Gujarat, Karnataka, Punjab, Haryana and Andhra Pradesh are relatively advanced, while Bihar, Uttar Pradesh, Madhya Pradesh and Odisha are relatively backward. So, what the country needs for continued economic growth, more equitable development and less poverty, is more productive agriculture, less corruption and more effective public institutions including, a freer press and a more rational legal system. In many of these areas, trends in the last 20 years show that India is worse off today than before.[19]

The reasons for this are a lack of decentralized reforms, a large but not modernized agricultural sector, abject rural poverty, mass illiteracy and the caste system. Within the federal structure, the role of the central government remains very strong, even after the economic reforms.[20] But the inherent nature of the government and state is still weak because of a corrupt bureaucracy and absence of effective governance measures, including rules and regulations, regulatory instruments and their networking throughout the nation – from the center to the states, cities, districts and local councils.[21] Moreover, the agricultural sector and rural areas, where about 70 percent of India lives, is neglected. One scholar writing on India's economic prospects observes, "For all its technological and industrial sophistication, India remains one of the most rural economies in the world, with the majority of its people dependent on jobs in an ailing agrarian sector that is incapable of absorbing the unemployed, let alone those about to become unemployed".[22]

This is a vast area where the concept of governance may be further developed, applied and practised by generating networks of self-governing rules among various groups of people for common purposes like irrigation, agricultural extension, rural healthcare, power generation and supply, rural transport, and educational infrastructure operation and maintenance. For this, the participation of local people, their training and education in various fields is also essential, and this can be undertaken by civil society groups in cooperation with the local government. This kind of space for governance is vastly open in other countries of South Asia.

So the nexus between the state, government, civil society and the people is very weak, and concomitantly, government accountability, transparency and people's participation in policymaking and implementation are lagging in South Asia. Much can be achieved by civil society organizations in collaboration with state agencies, particularly agriculture, education, community health, spread of information technology in rural areas, improving gender relations, development of women and children, human rights and alleviation of poverty. But the vital question that relates to the expansion of civil society's role in governance activities is the nature of civil society itself. Whether the civil society is democratic, undemocratic or violent, and may promote an authoritarian regime. "Democratic deficits within civil society jeopardize its ability to perform its proper social functions, and its legitimacy at home and abroad. Democracy requires not just more civil society, but better civil society".[23] There are widespread allegations in South Asian countries that civil society and non-governmental organizations (NGOs), which are also considered a part of civil society, are not necessarily democratic. Civil society organisations also lack accountability, transparency and tolerance in their behavioral and organisational affairs.

Regarding the federal structure of the state, despite constraints like the authoritarian, military rule in Pakistan, and strongly centralized bureaucracy in India, the scope of people's participation in governance in these countries is much greater than in highly centralized and unitary states such as Bangladesh and Sri Lanka. Unlike Bangladesh, although Sri Lanka has seven administrative provinces, the country is traditionally highly centralized. In fact, some argue that Sri Lanka's ethnic conflict could have been averted if the country had a federal political structure.[24] In India, since the late 1990s, there has been a growth in regional parties, along with the waning support base of the Congress, established by a highly centralized, nationwide organizational network. The regionalization of party politics has opened up opportunities for greater participation of people at the grassroots. Also, the growth of NGOs has strengthened people's participation, not only in activities like building schools, roads, houses and planting trees, but also to resist unilateral governmental decisions lacking popular support such as the Narmada dam project.[25] Despite all these, the government remains autonomous from the civil society because of the authoritarian nature of state, government and bureaucracy.[26]

There is a lack of transparency, accountability and people's participation in the government policies and activities in these states. Even in a democracy like India, there is limited scope for the people to make the government accountable and transparent, other than through the general elections. But except India, most countries in the region do not yet have laws granting access to information concerning the government or bureaucracy and public officials' actions. This is crucial for a vibrant civil society because, "access to information permits citizens to challenge government actions with which they disagree and to seek redress for official misconduct. Access to information also deters official misconduct by reminding public officials of their accountability".[27] In Bangladesh, colonial laws continue, preventing public access to information regarding government policies and actions even until very recently when the present government has passed a law granting the right to information.

The scope of the growth of civil society and NGOs or private voluntary organizations in India is much wider and eclectic because there are many social groups which are backward and socially segregated from the upper classes. These other backward classes (OBCs), including untouchables or outcaste Hindus, are amenable to organized efforts to improve their socio-economic conditions. This had provided political space for regional parties, first in the south then in the north, typically those with large backward caste populations. "India's democracy has become more inclusive and participatory",[28] writes Professor Ashutosh Varshney, citing several studies that support his observation.[29] In Pakistan and Bangladesh as well, public groups in diverse areas participate in socio-economic activities outside government institutions or protest government policies.

Democracy, Decentralization, Governance and Civil Society

As we have seen, South Asian states remain authoritarian and highly centralized. India and Pakistan, because of their federal structures, are decentralised to varied degrees, but Bangladesh being a unitary state, is strongly centralized, and so is Sri Lanka. Therefore, in India and Pakistan, although provincial/state governments provide space to the people for participation in governance activities, in Bangladesh there is no such space for the people to participate outside the central government. At the local government levels in all the countries, although there are district administration, *thana* or sub-district level administrative units, there is no elected government in these units. Although Bangladesh experimented with various types of representative local governments in different periods, none of them are in practice today at the two key administrative units, the sub-district and district levels. Nonetheless, the scope of civil society to organize people is enormous, to develop access to the government or to act independently. This is possible if the state is democratic and civil society can be organized on the democratic principles, as, "a broader strategy of governance that takes into account both building civil society and assisting political institutions is more likely to contribute to democracy".[30]

However, there are elected governments in the city and at the grassroots level, union parishad (council)/panchayat level local governments, where people may take part in elections, but not extensively in governance (in most South Asian countries except India). Thus, in Bangladesh, Pakistan and Sri Lanka, a vast space still exists between the central and local government for public participation by electing representatives (to govern at the minimum and for other governance matters at the maximum degree, for example building networks and connecting the people in collective and communitarian matters of decision-making and policy implementation).

Significantly, in Bangladesh, the 1997 Accord between the government and the Parbatya Chattagram Jana Samhati Samiti, the tribal people's organization of three Chittagong hill districts, provided for the setting up of three Hill District Councils and one Regional Council to be elected by the people. However, no elections to these bodies have been held till now. The question of decentralization for people's participation in governance affairs and for the development of self-governing regulatory rules at the local, regional or provincial levels is more pertinent in the case of Bangladesh, as Bangladesh is a unitary and highly centralized state.[31] One study observes, "Whatever the approach to the discourse on the issue of decentralized governance and development in Bangladesh, most scholars agree that local government institutions have not been able to emerge as 'self governing' units, but remained under the control of an all-powerful national government".[32]

Not only this, the civil bureaucracy that runs the government, including at the regional and local levels, is also highly centralized and traditionally resistant to reform, so the bureaucracy is insulated from accountability, transparency and people's control, and not conducive to good governance. This is more or less applicable to all South Asian nations, despite the fact that India and Pakistan have federal governments, and Bangladesh and Sri Lanka have a unitary form of government. There is no law even to ensure the people's right of access to information in these countries, except that India very recently enacted such a law by replacing the colonial Protection of Government Secrecy Act. The Freedom of Information Act is a pre-requisite to good governance, as it allows the people to know what the government and their representatives are doing, why and how. In democratic countries like the United States (US), there are many laws to ensure the people's right to information, including the federal Freedom of Information Act (all 50 states have statutes to ensure freedom of information), the Electronic Freedom of Information Act, 1996, the Government in the Sunshine Act, the Federal Advisory Committee Act, and the Ethics in Government Act. All these laws allow the people to know about their representatives, government functionaries and public policy matters without prior government permission.[33] But in South Asian countries, there is an enormous vacuum in all these areas, and specific laws still need to be enacted to support good governance with people's participation, establishing accountability and transparency. Worse, there is military rule in Pakistan and a civil war legacy in Sri Lanka.

Another barrier to good governance in South Asian countries is the authoritarian nature of the state and society. Traditionally, the state, political organizations and their leaders have tended to be extremely authoritarian in all South Asian countries.[34] Far from decentralizing power or promoting a democratic environment congenial for good governance, they have tried to concentrate more power in the hands of the states, so that no one can challenge their authority. But ironically, instead of making their positions stronger, these measures create situations where governance mechanisms remain weak and their roles in most cases rather ineffective (for example decisions or policies taken by the governments/states remain ineffective because of lack of proper implementation or execution).[35]

Another perception about the role of state in post-colonial societies such as Pakistan and Bangladesh, is that these states are autonomous,[36] and therefore they do not care much about people's participation in governance or accountability. Their autonomous role is possible because these states are relatively over-developed compared to the social and political institutions and social classes in these states of South Asia. That means state machineries are relatively stronger than the social and political institutions, for example, the political parties, political leaders and social groups.

Concept and State of Civil Society in South Asia

Civil society has been variously viewed, but the general consensus is that it comprises activities and organizations in the space that exists between the family and the state, which is "separate from and independent of state".[37] The modern concept of civil society is attributed to Hegel, who perceived it "as an intermediary sphere, the sum of all institutions that exist between the family, the most basic unit of social organisations, and the state, the most advanced".[38] This is a vast space that includes a range of organizations and social, cultural, economic and other activities. Alexis de Tocqueville referred to this kind of activity in the context of

American society as the great art of associating together to accomplish common works in society. He observed,

> In no country in the world has the principle of association been more successfully used or applied to a greater multitude of objects than in America. Besides the permanent associations which are established by law under the names of townships, cities, and countries, a vast number of others are formed and maintained by the agency of private individuals.[39]

Robert Putnam defines civil society as clubs and community associations, religious bodies, unions, professional associations, card parties, bowling leagues, bar cliques, ball games, picnics and parties, as well as voluntary organizations, social movements, and the internet.[40] However the concept of civil society is not yet very clear, and as an observer of American civil society writes, "When it comes to civic life, we are like visitors in a strange city without a map. The phrases 'civil society,' 'social capital' and 'civic virtue' sound as strange to most of us today as price/earnings ratio, ego and biodiversity must have sounded 50 years ago".[41]

The concept of civil society is not new in South Asia, even if its members may not be as educated or wealthy as in developed nations. India, Pakistan, Bangladesh and Sri Lanka have the tradition of organizing people for regional and local activities, institutions for building and running schools, colleges, forming elected local governments, running rural bazaars/markets, founding and managing religious institutions and cooperative societies in the agricultural, irrigation and rural economic sectors. One of the most significant and durable civil society institutions is *salish*, for alternative dispute resolution. Even in the colonial period, there were community-based organizations for promoting educational, religious, cultural and economic activity.[42]

But in the post-independence period, most of these societies have tended to become more dependent on the state, partly because of the influence of socialist ideology, and partly because of the rising expectations from the state of poverty-stricken people. One example of the transfer of responsibility from civil society organisations to the government is the transfer of a large number of high schools and colleges that were private institutions during

British and Pakistani rule in former east Pakistan, now Bangladesh. But it is now clear that state-centric development has been a failure in South Asia. On the other hand, the liberal, neoliberal market economy has also not been an efficient option, because of the lack of infrastructure, legal frameworks and regulatory mechanisms, among other things. Given this, there is a large space for civil society organizations in all South Asian nations, according to Fred Powell and Geoghegan.[43]

However, an outstanding phenomenon in contemporary South Asia and throughout the world is the mushrooming of NGOs and voluntary organizations. The question arises as to the effectiveness and democratization of civil society organizations in India, Bangladesh,[44] Sri Lanka and Pakistan, where a large number of NGOs, private voluntary organizations (PVOs) and semi-government organisations (SGOs) are active. In India, there are an estimated over one million registered NGOs,[45] while in Bangladesh there are over 47,000 NGOs. These NGOs work in wide-ranging areas, including, environment protection, empowerment of women, eradication of illiteracy, protection of human and child rights, alleviation of poverty, providing safe drinking water to people residing in rural areas and slums, and providing disaster relief. But "the management challenges and governance issues faced by Third Sector organizations are enormous and ill-understood".[46] If they are not democratic, that is, establish a self-governing, participatory policymaking mechanism, and ensure accountability and transparency in their actions and organizations, these CSOs may have their legitimacy questioned, even as they demand the same of the government.

There are allegations that many South Asian NGOs are not democratic, do not have participation of the stakeholders in the decision-making or policymaking processes, are not accountable to their constituencies and rarely transparent in their activities.[47] Most NGOs are accountable to their foreign donors. In Bangladesh, Akbar Ali Khan, former Chief of the Regulatory Reform Commission of Bangladesh (now defunct), a top bureaucrat and former advisor to the Caretaker Government of Bangladesh, recently observed, "most NGOs are not running for the benefit of the poor. Bureaucracy has gripped most of the NGOs".[48] The other CSOs, such as committees for schools, religious organizations, professional associations and even trade unions, are mostly elite associations, often have government representatives or are regulated by the government agencies in diverse ways. Nevertheless, since 1970s, NGOs and PVOs have been playing a crucial developmental role, by-passing the authority and role of the state and government to a great extent.[49]

Civil society may generate social and political capital to steer the political process of public policymaking and implementation of policies with people's participation and their voluntary actions.[50] If the civil society is developed, democracy would become sustainable where people would play a positive role in the governance process. "Even a democratic civil society does not ensure a democratic state, but the latter is unlikely to be sustainable without the former".[51] Building a democratic state is largely contingent on the construction of a democratic civil society. A civil society nurturing democratic ingredients such as pluralism, tolerance, competitiveness of social forces, participation by the people in communitarian affairs and generating civic pressure to ensure accountability and transparency in public affairs is crucial for building a democratic state.[52]

Some suggest that civil society organizations-state interaction, policies and regulatory measures are essential for growth and social development. The state can either directly support civil society to implement certain public policies/programs, or "not lead from the front, but to remove any hindrance".[53]

Nature of State-Civil Society Relations

State-civil society relations in South Asian countries fall in three categories – collaborative/co-operational, conflicting/confrontationist and disconnected or no relations with the state.

In the first category, organizations are developed either with state support or independently, to help the state implement its policies, for example, to reduce poverty, eradicate illiteracy, promote rural development, agricultural cooperative societies, population control, environment protection and tree plantation or "social forestry". Some NGOs therefore receive government funds to implement government programs, supported by donors.[54] In Bangladesh, it is observed that, "this refigures the NGOs, not as competitors to the state, but rather as potential new partners in mutually advantageous collaborative projects".[55]

In the second category, there are civil society organizations or activities developed to oppose state policies, for instance, development projects causing destruction of crops, cultivable land or environment, and forcing uprooting of people from their ancestral homes. In most cases, the human rights organizations protecting the rights of the marginalized have been suppressed by the dictatorial regimes in Pakistan and Bangladesh whenever they were or are in power, or by powerful local chieftains or government officials in all South Asian nations, including India. Authoritarian South Asian states tend to view any organization that questions state policy as anti-government or even, in extreme cases, anti-state. This has been more common during the military regimes in Pakistan and Bangladesh, the civil war in Sri Lanka, and the monarchy in Nepal.

The third category comprises CSOs or activities which are independent of government and its policies, for instance, *salish*, the alternative dispute resolution system in Bangladesh, for irrigation for small farmers, or management of social functions like marriages and deaths in rural areas.

As we have seen, there is vast scope for organizations autonomous of the government, and Galston cites four reasons for the "explosion of interest in civil society" in recent times.[56] Growth of civil society associations following the fall of the former Soviet Union and Soviet bloc countries; rise of NGOs worldwide protecting marginalized groups; growth of the idea of civil society, that is, voluntary organizations, as opposed to coercive state bodies; and concern in developed industrial societies over erosion of the traditional social fabric.[57]

In South Asia, the proliferation of NGOs, whether autonomous or collaborative, with the government, has not necessarily meant better prospects or a more vibrant civil society. They are often elitist, relations between the organizations and their stakeholders are hierarchical and non-participatory, and allegedly based on patron-client relationship. There are allegations of lack of transparency and accountability. A recent study by Transparency International Bangladesh (TIB) revealed that many NGOs were corrupt.[58] They have even been accused of being agents of foreign bodies, serving their donors, rather than their constituencies.[59] Therefore, the idea of civil society needs an overhaul and reorientation.

Challenges to Democratic Governance and Civil Society in South Asia

The challenges to build up a network of governance between government agencies, international organiszations, donors and civil society are manifold. These include poverty, corruption, traditional state structures such as a civil bureaucracy, military regimes, traditional business and the corporate world, traditional mindsets, low percentage of educated people and legal infrastructures, lack of transparency and accountability. The growth of democratic institutions such as a functional legislature, political parties, democratic leadership and replacement of colonial laws and machinery is essential to withstand these challenges.

While India is doing well compared to the other South Asian countries, Pakistan has repeatedly had military regimes, and "successive civilian administrations were unable to implement administrative and economic reforms."[60] So, Pakistan's governance travails viciously reinforce one another. Deepening poverty heightens the tensions between different ethnic and religious groups. Violence deters investment. Corruption, flight of capital, smuggling, gun-running, drug trafficking, and the evaporation of international donor and investor confidence, in turn, further undermine state capacity.[61]

Larry Diamond's observation on Pakistan in 2000 still hold true and the situation is rapidly becoming more precarious. Pakistan's prospect of development of democracy and a vibrant civil society is heavily undermined by the ever continuous coalition rule of the military and civil bureaucracy.[62] Bangladesh was under a state of Emergency for two years, 2007–08.

Given this, India remains a model for the rest of South Asia to identify the challenges and solutions to democracy, governance and building a vibrant civil society. Civil society organizations under authoritarian, opaque governments tend to be likewise. In the case of state-civil society collaborative relations, there is the danger of CSOs becoming agents of the state and promoting their interests, rather than those of the people.

Conclusion

The question of governance in the management of public and community affairs is of widespread concern today. Governments worldwide are considering or tackling reforms in their organizational, operational policy making and implementation measures. Good governance has become a social movement with civil society assuming ever greater responsibility worldwide. Consequently, governments have attempted reform on the one hand, and on the other, as NGOs – local, national and international–have proliferated, governments that have supported them.

The responsibility for initiating, developing and implementing rules for effective governance is shared both by the state and civil society. In South Asia, there is great scope for developing good governance by preparing civil society to undertake greater and more effective responsibility in community affairs. But it must be a good watchdog both of the government–and itself.

Notes and References

1. J. Pierre and B.G. Peters. 2000. *Governance, Politics and the State.* Basingstoke: Macmillan, p. 1, quoted in Andrew Jordan, R. K. W. Wurzel and Anthony Zito, "The Rise of 'New' Policy Instruments in Comparative Perspective: Has Governance Eclipsed Government?," *Political Studies*, 53: 477–96.
2. Anne Mette Kjaer. 2004. *Governance.* Cambridge, UK: Polity, pp. 1–2.
3. Ved Nanda. 2006. "The 'Good Governance' Concept Revisited". *ANNALS*, AAPSS, January. p. 269.
4. A.W. Rhode. 1997. *Understanding Governance, Policy, Networks, Governance, Reflexivity and Accountability.* Buckingham: Open University Press. p. 15.
5. Goran Hyden. "Governance and the Reconstitution of Political Order," in Richard Joseph (ed.), *State Conflict and Democracy in Africa*, Boulder, CO.

6. James N. Rosenau. 1995. "Governance in the Twenty-First Century," *Global Governance*, 1(1): 13–43.
7. D. Kaufmann, A. Kraay and M. Mastruzzi. 2005. *Governance Matters IV: Governance Indicators for 1996-2004*. Washington, D.C.: World Bank.
8. Center for Democracy and Governance. 1988. *Democracy and Governance: A Conceptual Framework*, Washington, D.C.: USAID, p. 19.
9. See Centre for Governance Studies, Bangladesh Rural Advancement Committee University, *The State of Governance in Bangladesh in 2006*, Dhaka: BRAC University, Centre for Governance Studies & BRAC. Research and Evaluation Division, 2006, Amiya Kumar Bagchi. 2005. "Governance-National, Corporate and International: The Indian Context," *South Asia: Journal of South Asian Studies*, 28(2): 265–81, Kamal Siddiqui, *Towards Good Governance in Bangladesh,* Dhaka; The University Press Ltd., 1996, Imtiaz Ahmed. 1999. "Governance and the international development community: Making sense of the Bangladesh experience," *Contemporary South Asia*, 8(3): 295–309.
10. See for the details on the emergency rule imposed by Indira government in 1976, Ramesh C. Thakur. 1976. "The Fate of India's Parliamentary Democracy," *Pacific Affairs*, 49(2): 276–93, Aaron S. Klieman. 1981. "Indira's India: Democracy and Crisis Government," *Political Science Quarterly,* 96(2): 241–59.
11. Katharine Adeney and Andrew Wyatt. 2004. "Democracy in South Asia: Getting beyond the Structure-Agency Dichotomy," *Political Studies*, 52(1): 1–18.
12. Johannes Jutting, Elena Corsi, Céline Kauffmann, Ida Mcdonnell, Holger Osterrieder, Nicolas Pinaud, and Lucia Wegner 2005. "What makes Decentralisation in Developing Countries Pro-poor?" *The European Journal of Development Research*, 17(4): 626–48.
13. See Atul Kohli. 1991. *Growing Discontent: India's Growing Crisis of Ungovernability.* Cambridge: Cambridge University Press, Javeed Alam. 1999. 'What is Happening Inside Indian Democracy?," *Economic and Political Weekly*, 34(37): 2649–56, and Stuart Corbridge. 2000. "Competing Inequalities: The Scheduled Tribes and the Reservations System in India's Jharkhand," *The Journal of Asian Studies,* 59(1): 62–85.
14. See Baldev Raj Nayar. 2003. "Globalisation and India's National Autonomy," *Commonwealth & Comparative Politics*, 41(2):1–16.
15. "The term 'Hindu rate of growth' was coined by Professor Raj Krishna to describe the relatively low rate of growth of India in the period 1950-80, when average GDP growth per annum was less than four percent," Marika Vicziany. 2005. "The Indian Economy in the Twenty-First Century: The Tough Questions That Just Won't Go Away," *South Asia: Journal of South Asian Studies*, 28(2): 211–32.

16. Gautam Adhikari. 2004. "Democratic Governance in South Asia," in Ramesh Thakur and Oddny Wiggen (eds), *South Asia in the World: Problem Solving Perspectives on Security, Sustainable Development and Governance,* Tokyo: United Nations University Press, Shrimal Perera, Michael Skully and J. Wickramanayake. 2005. "Human Progress in South Asia: A Multifaceted Analysis," *South Asia: Journal of South Asian Studies,* 28(3): 437–56.
17. See Christopher Lingle, "Indian 'New Deal' invokes bad, old idea," *The Japan Times,* February 9, 2006, Amiya Kumar Bagchi, en. 9.
18. See Patrick Heller. 2000. "Degrees of Democracy: Some Comparative Lessons from India," *World Politics,* 52(4): 484–519.
19. Marika Vicziany, "The Indian Economy in the Twenty-First Century: The Tough Questions That Just Won't Go Away," *South Asia: Journal of South Asian Studies,* 28(2): 211–32.
20. Marika Vicziany, 2005, pp. 222–28.
21. Peter R. Davis and Allister McGregor. "Civil Society, International Donors and Poverty in Bangladesh," *Commonwealth & Comparative Politics,* 38(1): 47–64.
22. See Marika Vicziany, 2005, pp. 223, and Raghav Gaiha and Vani Kulkarni. 1998. "Is Growth Central to Poverty Alleviation in India," *Journal of International Affairs,* 52(1).
23. Alison Brysk, "Democratising Civil Society in Latin America," *Journal of Democracy,* 11(3), 2000, p. 1.
24. See G. R. Tressie Leitan. 1998. "Federal Experience in Sri Lanka and Problems of Power Sharing," in Kouser J. Azam (ed.), *Federalism and Good Governance: Issues Across Cultures,* New Delhi: South Asian Publishers.
25. See Ahrar Ahmad, "The State, Participation, and Constitutionalism: Political Crises and Democracy in India," *Asian Affairs,* 26(1): 129–31, Ashutosh Varshney. 2000. "Is India Becoming More Democratic?," *The Journal of Asian Studies,* 59(1): 3–25.
26. For the concept of state autonomy in post-colonial countries such as those of South Asian nations see Hamza Alavi, "The State in Post-colonial Societies: Pakistan and Bangladesh," *New Left Review,* July–August, 1972, 59–81.
27. Robert G. Vaughan, "Transparency – the Mechanism: Open Government and Accountability," *Issues of Democracy: IIP Electronic Journals,* 5(2), August 2000, p. 13.
28. See Ashutosh Varshney, en. 20, 2000, p. 12.
29. See Ashutosh Varshney, citing the works of Dhirubhai Sheth, "The prospects of Pitfalls," *India Today,* August 31, 1996, Ashis Nandy, "Sustaining the Faith," *India Today,* August 31, 1996, Ashutosh Varshney, 1998, "Why Democracy Survives," *Journal of Democracy,* 9(3): 36–50, Myron Weiner, "Struggle for Equality," paper presented at the conference on "India at Fifty," October 31–November 1, 1997, Princeton University,

Princeton, NJ, Yogendra Yadav. 1996. "Reconfiguration of Indian Politics: State Assembly Elections, 1993–95," *Economic and Political Weekly*, (13–20 January): 95–104.
30. Nelson Kashfir, 1998, "Civil Society, the State and Democracy in Africa," *Commonwealth and Comparative Politics*, 36(2): 123–49.
31. See, for issue of decentralization in Bangladesh, Noor-e-Alam Siddiquee. 1997. *Decentralisation and Development: Theory and Practice in Bangladesh*, Dhaka: University of Dhaka.
32. Zerina Rahman Khan. 2000. "Decentralised Governance: Trials and Triumphs," in Rounaq Jahan (ed.), *Bangladesh Politics: Promise and Performance*, Dhaka; University Press Ltd., 2000 (2002), p. 109.
33. Robert G. Vaughan, *op cit.*, p. 16.
34. See Ayesha Jalal, *Democracy and Authoritarianism in South Asia*, Cambridge: Cambridge University Press, 1995.
35. What Davis and McGregor have commented on Bangladesh regarding the nature of a weak state and strong bureaucracy and military, may be said to be true in most of the South Asian countries, including India. See Peter R. Davis and Allister McGregor, *op cit.*
36. See Hamza Alavi. 1972. "The State in Post-colonial Societies: Pakistan and Bangladesh," *New Left Review*, July–August: pp. 59–81.
37. Gerard Clarke. 1998. *The Politics of NGOs in South East Asia*. London: Routledge, p. 9.
38. Gerard Clarke, Ibid. p. 10.
39. Alexis de Tocqueville. [1966] 1945. *Democracy in America*. New York: Alfred Knopf, volume I, p. 191.
40. Robert Putnam. 1999. *Bowling Alone: The Collapse and Revival of American Community*. New York: Simon & Schuster, p. 27.
41. Claire Gaudiani, "Our Ailing Civil Society," *Boston Globe*, 3 April 1996, quoted in Paul Rich, "American Voluntarism, Social Capital and Political Culture," *Annals of the American Academy of Political and Social Science*, 565 (September 1999): 15–34.
42. See Peter R. Davis and Allister McGregor, *op cit.*, p. 55, and John Staley. 1998. "Understanding Development: The Experience of Voluntary Agencies in South Asia," *Asian Affairs*, 29(II): 141–60.
43. Fred Powell and Martin Geoghegan. 2006. "Beyond Political Zoology: Community Development, Civil Society, and Strong Democracy," *Community Development Journal*, 41(2): 128–42.
44. One estimate says there are currently 47,000 NGOs in Bangladesh. See, "TIB Says its Study not Applicable to all NGOs: ADAB Decries Report,' *Daily Star*, October 7, 2007, pp. 1 & 15.
45. John Hailey. 1999. "Ladybirds, Missionaries and NGOs – Voluntary Organisations and Cooperatives in 50 Years of Development: A Historical Perspective on Future Challenges," *Public Administration and Development*, 19(467–485), p. 467.

46. John Hailey, Ibid, p. 469.
47. See Sarah C. White. 1999. 'NGOs, Civil Society, and the State in Bangladesh: The Politics of Representing the Poor," *Development and Change*, 30(2): 307-26.
48. 'Akbar questions most NGOs' transparency," *Daily Star*, November 7, 2007, pp. 1 & 15.
49. John Hailey, *op cit.*, Sarah C. White, *op cit.*, pp. 307-8.
50. Robert D. Putnam. 1993. *Making Democracy Work: Civic Tradition in Modern Italy*, Princeton: Princeton University Press.
51. Alison Brysk, 2000, *op. cit.*, (en. 20), p. 1.
52. See Fred Powell and Martin Geoghegan. 2006. "Beyond Political Zoology: Community Development, Civil Society, and Strong Democracy," *Community Development Journal*, 41(2): 128-42.
53. Sarah C. White, *op cit.*, p. 308.
54. See Hassan Zaman, et al. 2007. *Economics and Governance of Non-governmental Organisations in Bangladesh: World Bank Country Study*, Dhaka: The University Press Limited, pp. 64-68.
55. Sarah C. White, *op cit.*, p. 308.
56. William A. Galston. 2000. "Civil Society and the "Art of Association," *Journal of Democracy*, 11(1): 64-70.
57. Galston, 2000, pp. 64-65.
58. "TIB terms NGOs Mid-level Corrupt: Suggest Forming Independent NGO Commission," *Daily Star*, October 5, 2007, pp. 20 & 19.
59. See James Petras. 1999. "NGOs: In the Service of Imperialism," *Journal of Contemporary Asia*, 29 (4): 429-40.
60. Larry Diamond. 2000. "Is Pakistan the (Reverse) Wave of the Future," *Journal of Democracy*, 11 (3), p. 94, also see Marvin G. Weinbaum. 1996. "Civic Culture and Democracy in Pakistan", *Asian Survey*, 36(7): 639-54.
61. Diamond, Ibid.
62. Mahmood Monshipouri and Amjad Samuel. 1995. "Development and Democracy in Pakistan: Tenuous or Plausible Nexus?" *Asian Survey*, 35(11): 973-89.

PART I
NGOs in Collaborative Advocacy

2

Watershed Development Policies and Programs in India

Scope and Constraints of Civil Society Action

K. J. Joy

Overview

The way watershed development policies, programmes and practice have evolved in India over the last 20–25 years, is ample proof that civil society action, both in practice and policy advocacy, has played a significant role in shaping watershed development policies and programs. Having acknowledged this contribution, it is important to closely examine civil society action in the area of watershed development and ask critical questions like:

- Why is there greater scope for civil society action in watershed development, compared to, say, forestry?[1]
- What aspects of watershed development – bio-physical, social, institutional and property regime affect the scope and constraints of civil society action?
- What kinds of civil society organisations are involved in watershed development and engage with the state? This can impact the nature of its relationship with the state, as well as the agenda they may try to push through.

- Have the civil society organizations been able to bring new concerns to the watershed development policies and programs?
- What factors constrain civil society action? How can we redefine the role of civil society organizations, so that civil society can adopt a more "radical" agenda for watershed development policies and programs?

My attempt in this chapter is to analyse the scope and constraints of civil society action in the context of watershed development policies and programs in India, around the questions mentioned earlier. For this, I would primarily use four normative concerns – livelihoods, sustainability, equity and participation/democratization – the professed goals of watershed development. One of the core propositions of this paper is that the bio-physical, social, institutional and property regime characteristics of watershed development, as well as the nature of the civil society organizations involved, would have a bearing on both the scope and constraints of civil society action in watershed development.

The chapter is structured under six broad sections:

(a) Evolution of watershed programs in India.
(b) Bio-physical, social, institutional and property regime characteristics of watersheds.
(c) The four normative concerns of livelihoods, sustainability, equity and participation/democratization.
(d) Analysis of the kinds of civil society organizations involved in watershed development.
(e) Analysis of the scope and constraints of civil society action.
(f) Assessment of civil society action and some explanations why civil society action is what it is.

Evolution of Watershed Development Policies and Programs in India

In the 1970s, watershed development had no special significance for the development community in India. Some of the projects that became success stories, like Sukhomajri and Ralegaon Siddhi, were already under way, but received very little attention. However, the situation changed radically by the end of the 1980s. The major examples of watershed development taken up earlier left their mark. As a result, in the 1980s, the Indian Council of Agricultural Research (ICAR) initiated 42 "model watersheds" all over the country under the Operation Research Program.

The concept of integrated watershed development was first institutionalized in the National Watershed Development Programme for Rainfed Areas in 1990 with an allocation of INR 1,33,800 million in the Eighth Five-Year Plan. Following the Hanumantha Rao Committee's review in 1994 and the formulation of "Common Guidelines," the "first generation" projects implemented under the guidelines were in 1995–2001. We have also completed the "second generation" project phase supported by the Government of India under the Revised Guidelines of 2001, which have been further revised in 2003 as the Hariyali Guidelines.[2] The Ministry of Rural Development (MoRD) appointed a Technical Committee on Watershed Development in India, with A. Parthasarathy as chairperson, to assess the situation and suggest policy measures for improving efficacy of watershed development programs across different agro-ecological conditions in the country. The Committee submitted its report "From Hariyali to Neeranchal" in January 2006, and the government issued the "Common Guidelines" in 2008. These guidelines are binding and common to all watershed projects taken up under any government program/department. Thus the Common Guidelines mark a new phase in watershed development programs in India. Considerable work has also been done in the NGO sector, with support from bilateral and other donor agencies.

Watershed development is increasingly being seen as the lynchpin of rural development, especially in dryland areas; as something that would tie together and ground the various rural development efforts. Successful examples of watershed development seemed to offer a way out of stagnation and degradation for areas that economic development seemed to have otherwise bypassed, typically in drylands, semi-arid and arid regions perpetually under the shadow of drought.[3]

The country has also made significant investments using this approach.[4] By the end of the Eighth Five-Year Plan, an area of 4.23 million ha in about 2,554 watersheds had been treated and developed at an expenditure of ₹ INR 968 crore (GoI, 2001). In the Ninth Plan period, an outlay of INR 1,020 crore was provided to treat 2.25 million ha. Overall, including national funds and funds from bilateral, multilateral and private foreign donors, it is estimated that INR 2,400 crore (Farrington et al. 1999) have been spent annually on watershed development since the mid-1990s. Over the next 20–25 years, the target is to treat an area of 63 million ha, with an estimated total outlay of INR 76,000 crore (GoI 2000). The Parthasarathy Technical Committee Report also argues for greater fund allocation for watershed development in the country. According to the report, "for the (watershed) work to be completed by the year 2020, the government needs to allocate around INR 10,000 crore per annum for the next 15 years." (GoI 2006: 5). In the Eleventh Five-Year Plan, watershed development has acquired greater priority, and per ha allocation has been nearly doubled, from INR 6,000/ha to about INR 12,000/ha. This is also reflected in the 2008 Common Guidelines.

Thus, watershed development is no longer seen as an "experiment," but viewed by everybody – governments, donors and non-governmental organizations (NGOs) – as a core strategy to stabilize rural livelihoods through its multi-sectoral approach, especially in the dry, rainfed regions of India.

Of course, the performance of watershed development programs has been rather mixed. Though there have been many positive examples of watershed development in the country that demonstrate its potential, the average performance across different modes of watershed development has been rather mixed and often below expectation.[5]

Characteristics and Property Regimes of Watersheds

A watershed is a natural geo-hydrological unit of land, which collects rainwater and drains it through a common exit point. Such a unit can be as small as a few hectares of a first order stream, to as large as thousands of square kilometers such as that of a river basin. In India the watershed development programs are pitched at an average scale of about 500 to 1000 ha.[6]

A watershed comprises three broad zones – recharge zone, transition zone and discharge zone. Sometimes they are also called upper reach, middle reach and lower reach, and each has its own bio-physical characteristics. For example, the recharge zone could be of higher slopes, with very little soil cover. On the other hand, the discharge zone is much plainer with better soil conditions. Depending on the slope and soil conditions, watersheds are also sub-divided into various land capability classes.[7] One of the cardinal principles of watershed development is that watershed interventions and land use must follow the land capability classification. Most of the soil and water conservation works take place in the upper reaches; whereas most of the benefits of these works, in terms of recharge, show up in the lower valley area. In the absence of institutional arrangements, people with land in the lower area derive most of the benefits of water recharge.

Watershed development also has social and administrative boundaries. It is also the space of livelihood for most rural people. Thus it is important to understand the interconnectedness of the bio-physical and the social spaces while analysing watershed development. Historically, social factors, property regimes, resource use, that is, the interaction between the cultural, social and economic spheres, impact watershed development; it is not a static entity, but ever evolving.

Watershed development is primarily a land-based program, consequently the bulk of the benefits go to those who own land and that too in proportion to the size of their holdings. Thus, property right in land is a pre-requisite to get benefits from a watershed program.[8]

There are two types of inequities that impact who benefits from watershed development programs. The first is historic inequalities on the basis of class, caste, patriarchy, ethnicity, and so on.[9] Thus to correctly assess the impact of watershed development, one needs to disaggregate the local community in terms of different social sections (class, caste, ethnicity, etc.) and their differential access to productive resources. Patriarchy is another issue, and there could be differential impacts on women within households.

The second dimension emanates from spatial or locational inequalities, primarily because of the bio-physical characteristics of the watershed itself. One's location in the watershed often determines one's access to recharged water. As we saw, people who own land in the valley portion benefit most from the augmented water resource.

It is important to understand how the locational inequities map on to the historically embedded inequities of access to productive resources and what impact watershed development has on them. Generally, the locational inequities map on to the historically embedded inequities and the combination is such that the inequalities worsen. This is because (*a*) land in the upper reaches is relatively more often owned by the poor, and the lower reaches by the rich and upper castes, (*b*) watershed development augments groundwater, which is currently private property, and can be tapped much better by the rich and the landed, than by the poor, (*c*) normally, increased availability or assurance of water does not directly benefit the landless in any case.

In India, surface water is generally considered common property, and groundwater private property. It is a paradox that through watershed development measures, we are also changing the property regime of the water – when surface flow, which is in the common property regime, appears as groundwater, it becomes part of the private property regime. And obviously this has implications for who gets access to the recharged water.

Finally, the watershed development program is supposed to operate under "consensus" and many community-based organisations (CBOs) like a watershed committee, area groups or user groups, are supposed to be set up as part of the program implementation design. The process of building up consensus pre-supposes that one needs to go beyond class and caste divisions and polarizations within the community, and build institutions overarching these divisions.

Livelihood, Sustainability, Equity and Participation in Watersheds

Over the past three decades, the concepts that determine the goals of watershed development programs, have also evolved along with the content of the programs. Catchment protection programs, and soil and water conservation programs were the precursors of watershed development. Catchment protection programs looked upon the watershed as a unit, but they focused on the catchments of particular dams, mainly to reduce sediment load and siltation of the reservoir. Soil conservation programs aimed at conserving fertile agricultural soil through bunding; but the bunding considered the farmer's field as a unit and lacked any larger unit of organization. Check dams and other waterline treatments carried out for water conservation were taken up in a standalone manner, without being integrated into a watershed-scale program.

With the emergence of watershed development as a distinct program, soil and water conservation acquired a unit of organization – the watershed. Soil and water conservation are still central to watershed development, and afforestation, common land regeneration and agronomic changes are linked to this central theme. Of late, watershed development is being increasingly seen as a core strategy for stabilizing rural livelihoods, especially in the dry, rainfed regions of India by everyone concerned – governments, donors and NGOs. All other developmental issues, including employment generation programs, rural credit, women's empowerment – and even prohibition and population control as in the case of Adarsh Gaon Yojana in Maharashtra – are being subsumed under this concept.[10]

The goals of watershed development have shifted to include participation, gender equity, sustainability and livelihoods.

These are increasingly reflected in the Common Guidelines of 1995, the Revised Guidelines of 2001 and Common Guidelines of 2008.[11] For example, the Common Principles for Watershed Development aim to promote equity for the resource poor and women, and suggest that "equitable right to all households in any new water resources developed under the project" as one of the ways to achieve this (MANAGE 2000).[12] The most extreme example of this shift is that of Karnataka Watershed Development Society (KAWAD), which prefers to call its program a "livelihood program with a watershed approach." The Common Guidelines, 2008, also emphasise livelihood, equity and gender. It says "watershed development projects should be considered levers of inclusiveness; project implementing agencies must facilitate the equity processes such as enhanced livelihood opportunities for the poor through investment in their assets and improvements in productivity and income; improving access of the poor, especially women, to the benefits; enhancing the role of women in decision-making processes and their representation in the institutional arrangements; ensuring access to usufruct rights, that is the right to use the products or produce, from the common property resources for the resource poor" (GoI 2008).

Here, I briefly analyse the four important normative concerns – livelihoods, sustainability, equity and participation/democratization underpinning watershed programs, at least at the level of rhetoric.[13]

Livelihoods

A watershed development program is supposed to meet the livelihood needs of the people. Livelihoods broadly include domestic water (water for drinking, cooking, washing, hygiene and livestock), food, fuel, fodder, some biomass inputs to maintain soil productivity, as well as other goods and services that may have to be obtained from the larger system (health, education, entertainment, transport, communication, etc.). Apart from meeting these needs, it also includes needs arising from the livelihood activity itself. So a livelihood comprises the capabilities, assets and activities required for a means of living. Livelihoods cannot be equated with (cash) income, as an increase in income does not necessarily mean that it contributes to livelihoods.[14] The overall objective is to be increasingly self-reliant – in the case of needs like food, fodder, fuel and domestic water, self sufficiency *in kind* is the aim; the other needs can be mediated by cash through equal exchange.

Sustainability

Though sustainability means different things to different people, in the context of watershed development it refers to sustainability of the underlying bio-physical processes, their environmental integrity and dependability as mediated by human intervention. Practical measures for sustainability could include:

(a) Conservation and/or enhancement of the primary productive and assimilative potential of the ecosystem.
(b) Use of resources like water and biomass within their annual flows or within renewable limits, without drawing from stocks. Stocks to be used only in bad years with the understanding that they would be replenished in good years.
(c) The rate of regeneration of the resource must be greater than or equal to the rate of harvest.
(d) Minimization of the import of water; and if needed in exceptional situations, it must be done in a fair manner.
(e) Sustained productivity of agricultural and common lands (including low-input, low-impact agricultural practices).
(f) Ensuring sustainability of downstream agro-ecosystems.
(g) Increasing diversity of crops and crop practices, and agro-ecological processes.
(h) Shift from high input based agriculture to more sustainable practices like Low External Input-Based Sustainable Agriculture (LEISA).
(i) Increased use of local, renewable and energy efficient materials, especially in construction of structures.
(j) Plan for greater dependability and assurance.

Equity

Equity in the context of watershed development, means more equitable access to the benefits of watershed development programs, taking into account both historic and locational disadvantages. It means more equitable access to natural resources, especially the augmented resources (like the increased water and biomass) generated by the watershed development programs. The avenues for equity could include the following:

(a) Privileged access for the resource poor in common property land resources.

(b) Positive discrimination, that is, favoring those who are disadvantaged with respect to class, caste, ethnicity, gender and location, and providing women with preferential access to water, both for domestic and productive uses.
(c) Inter-sectoral equity and water use prioritization: water for domestic use getting an overriding priority over other uses.
(d) Equal opportunity for participation in decision-making, management and governance functions, especially in the institutions.
(e) Enhancing opportunities for resource poor sections through non-farm livelihood activities.

Participation and Democratization

Participation needs to be seen both a means and an end. So participation is a means to bring in, efficiency, equity and sustainability, and is also a goal in itself. Very often, participation is equated with voluntary labor (*shramdan*) and participatory rural appraisal. If participation has to lead to democratization or empowerment, it should include the following:

(a) Informed choices made by the community.
(b) Primacy of the local community[15] in decision-making.
(c) More control for local communities in the design and implementation of the watershed development programs.
(d) Accountability of larger structures and agents (supra local agencies) to the local community ("downward accountability").
(e) Separation of allocation and regulation/governance functions from service delivery or production-related functions.
(f) Representation of women, landless and other resource poor and marginalized sections in the institutions and decision-making processes.
(g) Right to information, data and experiences from other areas.
(h) Financial transparency.

(*i*) Outsiders having a role in capability building of the local communities to make informed choices, and in raising issues of equity and sustainability.

(*j*) The relationship between the local communities and "outsiders" should be a two-way traffic, with learning for both.

In a way, the preceding discussion of the four normative concerns of livelihoods, sustainability, equity and participation/democratization is more like the Weberian "ideal types". In the real world of watershed development, conditions are not ideal; they are mediated and contested through existing inequitable and exploitative social relations (class, caste, patriarchy, ethnicity, etc). The limited manner in which these concerns have been addressed in watershed development, if at all, is testimony that watershed development is a highly contested terrain. I return to these concerns and in what manner they have been addressed or not in Section five, where I discuss the scope and constraints of civil society action.

Watershed Development and Civil Society Organizations

To unravel the scope and constraints of civil society action in watershed development, it is very important to critically examine the type of civil society organizations engaged in the watershed development sector. It is premised that the type of civil society organizations involved will affect the nature and scope of civil society action. This study is limited to identifying the types of civil society organizations in watershed development. Understanding their distinguishing features would make issues clearer and set the boundaries of their engagement.

Often, civil society organisations[16] are equated with NGOs which could include community based development organizations, social welfare organizations, social action groups, grassroots organizations, etc. These may differ greatly in defining the problems and the strategies they adopt to address the problems. Welfare-service type charitable organizations are set up primarily by religious, secular civic bodies or philanthropists to offer medical services, housing, food, day-care facilities, etc. Community development organizations implement social and economic development programs (e.g. programs related to agricultural development, natural resource management, health and education, income-generation, etc.). Social action groups represent the issues of marginalized people through lobbying, pressure group strategies and advocacy.

Within the social work discourse, Jack Rothman et al. (1987) talk of three models of community organisation – locality development, social planning and social action. Locality development focuses on the common needs of the locals, with some local participation. In social planning, an expert or a planner advises the community how to resolve its problems. Social action takes a specific position on behalf of the underprivileged and works towards a more equitable power structure and re-distribution of resources, with the community organizer or activist/advocate playing a key role (Rothman et al. 1987). Social action groups represent voluntary effort within NGOs, or the left discourse of non-party political formations.[17]

Swapan Garain (1994) says there are three types of NGOs:

(a) Corporate NGOs which are heavily government sponsored (such as the National Dairy Development Board, the Wasteland Development Board) and have marginalized the issues of the poor to promote capitalist development;
(b) Development-oriented NGOs which rely on government patronage and hence do not question government policy and programs and further the state's capitalist interests;
(c) NGOs that aim for development with social justice and promote the rights of the poor, are often in conflict with the state and the elite (Garain 1994 as cited in Kamat 2002: 14). This category is pretty close to the social action groups mentioned earlier.

Sometimes non-party political formations and new social movements are also clubbed with social action groups or seen as part of broader citizens' action groups. They are mass based organizations working primarily among different marginalised social sections, use agitational methods, take political positions on national and international events and address issues including, class, caste, patriarchy, ethnicity, minority and newer concerns like environmental sustainability. They aim for transformation and, by and large, do not rely on institutional funding, especially foreign funding. New social movements bring relatively newer issues to the political agenda, including environmental issues, struggles against big dams and displacement, movements against pollution and for the drought-affected to have equitable access to water.

The defining characteristics of new social movements, as discussed by Omvedt (1993) are, "They are 'social movements' in the sense of having a broad overall organisation, structure and ideology aiming at social change."

They are "new" in that they themselves, through the ideologies they generate, define their exploitation and oppression in "new" terms – related to traditional Marxism, but having clear differences with it. They cannot, in other words, be seen as simply "popular movements" willing to follow the leadership of the working class, its parties or ideology; they consciously reject this kind of relationship and question its ideology.

They are movements of groups that were either ignored or exploited by traditional Marxism (women, Dalits and shudras) or who are exploited by contemporary capitalism (peasants forced to produce for capital through market exploitation managed by the state; peasants and forest dwellers victimized by environmental degradation) but overlooked through a preoccupation with the exploitation or surplus extraction associated with "private property" and wage labour.

A proper analysis of their position calls for a modified Marxist analysis – or historical materialist analysis – of contemporary capitalism (Omvedt 1993: xv–xvi). In fact, these defining characteristics of new social movements distinguish them from typical social action groups or other welfare/development oriented voluntary organizations.

Civil society organisations can also be classified into two categories on the issue of how they relate to the state: those that work within the space provided by the state and do not challenge existing social relations of production, and those that challenge them with a transformative agenda. In fact, this has become all the more critical with globalization and economic reforms since the early 1990s. In fact community based organizations find themselves in a peculiarly compromised position, given that international agencies and the state are encouraging them to take over some of the public services that were previously rendered by the state (Bhatnagar and Williams 1992 as cited in Kamat 2002: 9). The rationale advanced is that the mandate of the NGOs is not very different from the logic of neoliberalism; that both the market and the "popular sectors" need to develop their own local initiatives, and hence, "the state should be less involved in intervention, and civil society (including the NGOs) more so" (Bebbington et al. 1993 as cited in Kamat 2002: 9). However, the authors (representing the Overseas Development Institute) are careful to allocate this new role to only "private non-profit institutions, both national and international, which provide direct services to farmers, but are not themselves farmer organisations" (ibid.). Thus, the attempt is to legitimise only those that implement and initiate development projects, and which involve professionals, leaving out organizations that are based on mass membership (Kamat 2002).

This is also very much related to the general process of "NGOisation" of the "development" sector that began in the 1980s. Development-oriented NGOs seem to be increasingly occupying the space left by the 'receding state' in terms of service delivery, and also the political space that was once occupied by explicit political movements, both party and non-party. This has led to the "NGOization" of the development sector and impacted the development discourse.

Civil society organizations involved in the watershed sector are primarily NGOs (community development organizations or development oriented NGOs as per Garain's classification), or what are called "private non-profit institutions", registered either as a trust or society, or as both. Some of them are registered under the Foreign Currency Regulation Act (FCRA) to access foreign funding, and are often professionally managed. Some with the clout to access financial resources are also engaged in watershed development. Some not only implement watershed development projects by accessing both government, as well as foreign and Indian donor funding, but are also involved in training, capacity building and policy advocacy. The others are involved in research and documentation. The other civil society organizations like mass-based groups, social action groups or non-party political formations and new social movements are not involved in watershed development sector. In other words, the civil society organizations involved in watershed development are mainly those who want to work within the space provided by the state or within the limits imposed by the present social relations. Of course, there are differences between these civil society organizations; but those differences are within the boundary conditions set by the broad type of categorization of CSOs. To this we should also add the donor agencies, including bilateral and multilateral agencies like the World Bank, which also significantly impact watershed policies and programs. They not only provide funds, but the funds often come with certain conditions and guidelines.

What Civil Society Action Can Achieve

My attempt here is to address civil society action at two levels. First, let us examine the role of civil society action in making watershed based development, at least at the level of rhetoric, one of the core strategies for rural development. Second, let us see to what extent civil society action has been able to bring in relatively newer concerns of livelihoods, sustainability, equity and participation/democratization into watershed discourse – both in policy and practice.

After Independence, the development of agriculture, and rural development itself, was primarily based on providing irrigation water, that too from centralized, large irrigation projects, often described as "the hydraulic mission"[18] – a legacy of British times. Dams became the "temples of development." This approach was further strengthened by the Green Revolution, ushered in India since the 1960s, when water, hybrid seeds, fertilizers, pesticides and credit became the main drivers of agricultural growth. Such a strategy left vast areas untouched and people in these areas – mainly the drought prone regions and the hilly tracts – to fend for themselves through rainfed agriculture. Though the Green Revolution strategy did contribute to creating food "surplus" by pushing up agricultural production, it also created islands of prosperity, giving rise to the great debate in the 1970s and 1980s of "uneven development of capital." However, by the 1990s, as productivity declined, it became increasingly clear that it would be difficult to sustain such a strategy economically and ecologically.

The 1980s and 1990s also saw a large number of social movements against large dams in almost all parts of India, the most prominent one being the Narmada Bachao Andolan (NBA). Though it started as a movement against the submergence and displacement of the Sardar Sarovar Project (SSP) on the Narmada, it soon became an anti-dam agitation and raised the issue of "large v/s small" in the development discourse. The highlight of the movement was when the World Bank withdrew from the SSP in 1993, following the Morse Report.[19] Further, with the publication of *Dams and Development: A New Framework for Decision Making* by the World Commission on Dams in 2000, there was an increasing realization that large dams need to be seen as one of many options, and not the only option, for water resource development (World Commission on Dams 2000).[20] The National Alliance of People's Movements (NAPM) provided a national, political platform for a broad spectrum of civil society organizations, ranging from mass-based groups like NBA and the Fish Workers' Union to typical NGOs in the 1990s, to fight displacement and other forms of "destructive development".

Another social movement trying to engage with the state through an alternative agenda for drought eradication and water resource development and distribution was the erstwhile Mukti Sangharsh Movement (MSM), now part of the wider Pani Sangharsh Chalwal in south Maharashtra. In the 1980s, when watershed development as a concept and strategy had not gained much currency, MSM forced the Government of Maharashtra to take up water and soil conservation works like bunding, percolation tanks and check dams under the Employment Guarantee Scheme (EGS) in the drought prone areas of Sangli district. In fact, under the leadership of MSM, the drought affected people of Khanapur taluka (Sangli district) drew up alternative plans, showed the district authorities the sites where such works could be taken up, and organized struggles to get them implemented. Later the Pani Sangharsh Chalwal started mobilizing people in south Maharashtra to demand restructuring of the large lift irrigation schemes on the river Krishna. The main demands have been: integration of local and exogenous water sources – local water harvesting through extensive watershed development and water from the river Krishna – and equitable distribution so that each household got enough water for its livelihood needs.[21] Unlike NBA, the Pani Sangharsh Chalwal is not against dams, but argues for the integration of large and small water sources.

These are two mass based CSOs (or new social movements) trying to engage with the state on its water related agenda. Of course, there are many more organizations who have tried to impress upon the state and society at large, the need for more decentralised and integrated water resource development.

The 1980s also saw some successful experiments in watershed-based, decentralized development, of which Sukhomajri and Ralegaon are the most well known. These and many other successful experiments in watershed-based development demonstrated the potential of a watershed-based approach to address both ecological degradation and assurance of livelihoods in rain fed areas.

Over the last decade or so, there has been a partial shift from the thinking that centralised, large irrigation projects are the only answers to the problems of degraded, dry lands of India. In fact, the National Water Policy (2002) and state level water policies at least talk about the need for watershed development. This is also one of the reasons for the increasing importance and growing financial allocations for the watershed development program in the country.[22]

This shift in thinking has been due to many reasons – from the crisis in the irrigation sector; the Green Revolution reaching a plateau; disillusionment with top down, centralized, bureaucratic practices; the global discourse on sustainable, decentralized and participatory paradigms, and so on. The social movements against large dams (especially NBA and others) did help in changing mindsets and seriously considering watershed-based, decentralized, local systems as an option. Here, the primary contribution has been from CSOs which we have characterized as mass based, non-party political formations or new social movements. Of course, the innovative and successful experiments in watershed development helped in demonstrating the potential of this approach. These struggles and critiques against big dams have created an ideological space, conducive for a decentralized, participatory, watershed-based development approach.

Let us examine to what extent civil society action has promoted the four normative concerns of livelihoods, sustainability, equity and participation/democratization and how they impact the actual content of watershed development programs.

NGOs and Livelihoods

Though watershed development projects are increasingly seen as livelihood projects or are expected to contribute to livelihood security, one cannot clearly say that this increased concern is primarily due to civil society action. Of course, some leading NGOs in watershed implementation have definitely experimented with different approaches, especially taking up a more integrated farming system approach, combining agriculture with horticulture, silviculture (tree cultivation), pastures, livestock, etc., to diversify livelihood opportunities. For example, the Bharatiya Agro Industries Foundation (BAIF) tried to experiment with the *wadi* (small orchard) program, which has now become part of many mainstream watershed programs. The non-land based livelihood component, which has been added to the new programs with clear budgetary provisions, owes its origins primarily to the work of NGOs. Otherwise watershed development has always remained a land-based program.

Various reviews of watershed development programs have shown that the issue of livelihood has not been addressed comprehensively (or as defined in Section Three earlier). There is no clear understanding or consensus as to what watershed can or cannot do in the area of livelihoods as minimum benchmarks. Many NGOs involved in watershed development programs still do not believe that watershed development should be able to meet basic needs like domestic water, fuel and fodder, and there is no conscious effort to make this part of the program design.[23] Moreover, the fact is that watershed development has had a better impact in years with good rainfall, but not in years with drought or below average rainfall. The main lacuna has been that dependability of rainfall has not been taken into account while doing the watershed planning. Even in some of the NGO-implemented watersheds, though considered successful projects, drinking water had to be supplied through water tankers in drought years.[24]

Though present-generation watershed development programs have non-land based income generation activities for the landless and other resource poor sections, mostly these are the same old conventional activities like pickle or *papad* making. These activities often suffer from market saturation and they are also not linked to the core of the watershed program. If the PIAs are innovative then they could think of many other ways of enhancing the livelihood avenues of the people through activities, which are directly related to watershed development. For example, if part of the bulk biomass (small dimension timber, bamboo, fiber, etc.) produced as a result of watershed development is made available to the resource poor sections, they can improve their livelihood by processing it, adding value and using it for infrastructure, for example, water harvesting, buildings, roads, etc.[25]

CSOs and Sustainability

Civil society action does not seem to have contributed much in redefining the term sustainability as we have defined in Section Three. Though there is an increasing awareness about the need to regulate groundwater extraction, it has not been translated into specific policy prescriptions or actions on the ground, except may be in very few cases as in Hivre Bazaar in Maharashtra. Often, even NGO-implemented watershed projects are preoccupied with bringing more and more non-agricultural land under agriculture, without paying due attention to the specific ecological niche of such non-agricultural lands, and continue to adopt Green Revolution agronomical practices for productivity enhancement.

NGOs and Equity

Almost all watershed guidelines include equity and fairness in distribution of the benefits from watershed development. Many NGOs, including larger ones like the Mysore Resettlement and

Development Agency (MYRADA) showed better results after promoting Self Help Groups (SHGs) among resource-poor sections, especially women. They helped women save, obtain credit, and become more active and visible in community affairs. Self-help groups have been included in policy and are now mandatory in all watershed development guidelines. But this activity has not become an integral part of the watershed development and has had little impact on traditional gender roles.

Civil society action has also not been able to do much in making clear policy prescriptions regarding sharing of the increased resources. In fact, most NGOs do not make any effort to ensure equitable sharing of increased resources like water or biomass, but believe that the primary benefit for resource-poor sections is greater wage employment in itself. They fear that talk of equity would result in conflicts within the community, and prefer to avoid addressing it.

Some NGOs believe that equity is an agenda imposed from outside, but the same NGOs have no qualms in accepting money from outside or agreeing to conditions imposed by external funding agencies regarding, for example, the extent of the people's contribution or imposing a ban on grazing.

There are instances where a higher incidence of landlessness has made a village ineligible for a watershed development program. For example, the Indo-German Watershed Development Program gives preference to villages with fewer landless people, and the MYRADA also has a similar criterion. This is a frank admission by the implementing agencies of the limitations of the watershed program in addressing the livelihood issues of the resource-poor sections, especially where the proportion of the landless is very high (Joy and Paranjape 2004).

Nor have many NGO-implemented projects show sufficient sensitivity to the historical and locational inequities discussed in Section Three (ibid.). In this context, it must be noted that the term "equity" is increasingly replacing good, old-fashioned "equality", that was used to denote fairness in the distribution of and access to resources, till the early 1980s. Equality has been inscribed on the banner of all radical movements for social change.

These movements perceive inequality as the result of the social structure, so the demand for equality has usually called for a radical, egalitarian social transformation, with structural changes in society. However, since the 1980s, for a host of reasons including globalization, economic reforms, growth of the voluntary sector and growing "NGOization," equality, which still smacked of an earlier radicalism, has been increasingly replaced by equity. With this, the emphasis also shifted from the "radical projects" that characterized the radical mass movements to what is immediately possible and practical. Thus, in the context of watershed development, equity is preferred to equality, because people are usually talking about what can be done *without* a radical social restructuring, preferring to work within limits of the existing social relations. In other words, it points out and demarcates the space available within the present system. This means that many agencies are content to reduce equity by creating, for example, preferential access (not necessarily ownership) for small parcels of land and providing limited quantities of water to disadvantaged sections, while genuine equality remains a distant dream, if at all.[26]

Although it is true that the asymmetric nature of watershed processes makes watershed development 'naturally' prone to aggravating intra- and inter-village/watershed inequities,[27] it must be noted that watershed development has immense inherent potential for equity, though it tends to be realized only through strong, proactive initiatives. Watershed development enhances ecosystem resources and productive potential. Moreover, this enhancement usually takes place through public funds and collective, community effort. Thus it can be argued, that *the additional resources that have been created through watershed development must be distributed equitably to everyone in the watershed, even as rights to previously existing resources are recognised and left largely undisturbed.* Thus, watershed development can be a positive sum game, creating potential access to productive resources for the rural poor, even without greatly disturbing prior rights and use. Unfortunately, few NGOs seem aware of the true potential of watershed development, even within the limited framework of equity.

Making Watersheds Participatory

The biggest contribution of civil society action is making watershed programmes more participatory. The nomenclature itself changed from "watershed development" to "participatory watershed development,"[28] and the emphasis shifted from a techno-centric approach to a participatory one, with social and institutional issues being given due importance. Indeed, public participation is an important component of all watershed program guidelines, especially since the 1995 guidelines following the Hanumantha Rao Committee report, including those from the Ministry of Agriculture for the National Watershed Development Programme for Rain fed Areas. Various options are suggested to promote public participation, through community mobilization, participatory rural appraisals and institutions like gram sabhas, SHGs, user groups and watershed development committees. Often a distinction is made between the "hard" and "soft" components of the program: the hard components refer to technical, bio-physical interventions that are planned, designed and implemented by the line departments, whereas the soft components refer to community mobilisation, participatory rural appraisals, exposure visits, training, etc., which are usually handled by the NGOs. The criteria for success usually include not only physical targets, but also indicators like public participation.

The Watershed Support Services and Activities Network (WASSAN, Hyderabad) and its network of NGOs have not only worked out detailed process indicators internally, they even did advocacy to convince the Andhra Pradesh government to adopt similar process guidelines.

It is important to allocate time and resources appropriately for institutional and capacity building of the watershed community. In the Indo-German Watershed Programme, there is a six months to one year preparatory phase, in which the main emphasis is on enlisting peoples' participation through *shramdan*, and also building the capacity of the community through exposure visits, training and demonstration. The Forum for Watershed Research and Policy Dialogue (ForWaRD)[29] recommended to the Parthasarathy Technical Committee Report on Watersheds, Ministry of Rural Development, 2006, that watershed programs operate in phases, and emphasized the preparatory phase, building institutions and the capacity of the people, allowing them sufficient time to demonstrate their willingness to participate in the program.[30] The Common Guidelines of the Government of India, 2008 also include this point.

Though there is a perception that participation is often unambiguously endorsed on normative grounds, even if the empirical basis is not clear (Cohen and Uphoff 1980), there is sufficient empirical evidence to say that participation is a necessary, though not sufficient condition, for a better outcome. Many studies have shown that NGO – run projects had better outcomes, mainly because of their emphasis on participatory processes (Kerr et al. 1999).

Though civil society action emphasising public participation resulted in greater awareness and was implemented in policy and practice, it is mostly viewed as a means to obtain cooperation, raise efficiency, and gain legitimacy, rather than an empowering objective in itself. Most of the decision-making is still made by the project implementing agency. Very often the participation is limited to *shramdan*, participatory rural appraisal, operation and maintenance of structures, and so on. However, participation of the local community in crucial decision-making and control of fund allocation and expenditure, has been pretty dismal.

Major decisions are usually taken beforehand by the project implementing agency, and consultation with the local community is usually with the powerful in the community. There have not been enough serious efforts to address inequities, hierarchies and power relations within the community, and there is also the danger of "elite capture" of the participatory institutions.

Most guidelines stress cost sharing as an indicator of participation, with some NGOs arguing for greater contribution from the local community. But insistence on cost-sharing as a component of public participation can bring its own distortions. For example, MYRADA recommends that there should not be any subsidy component in the program, meaning 100 percent cost recovery. This brings the danger that the poor may be "priced out" of the program because they cannot afford the contribution. Also, the principle of "ridge to valley" – one of the cardinal principles of watershed development – is not always followed because the program is based on people's willingness to pay. Sometimes, contributions come from wages – either deducted or withheld. Effectively, this means that the poor pay on behalf of the landed as it is the resource poor who provide the necessary labour on watershed development projects.

Similarly, participatory rural appraisals – primarily championed by NGOs – can also be distorted. It is being increasingly used as a tool for data collection, to enlist local participation and indicate local development priorities. Even when not reduced to a bureaucratic procedure, it is problematic because often it may represent only the opinion of a few, especially, dominant sections in the village. It is important to contextualize participatory rural appraisals and be aware of its possibilities and limitations. It can be an effective tool for a qualitative and rapid understanding of the situation. However, it does not provide reliable quantitative data regarding resource status or land use patterns, and may leave no room for interactive learning between local knowledge systems and external, modern systems of knowledge.

Though the 73rd constitutional amendment on Panchayati Raj was a decisive step towards political, administrative and economic decentralization, there is not enough space for Panchayati Raj Institutions (PRIs), and the relationship between them and watershed development organizations, especially NGOs, have often been problematic. By and large, NGOs prefer to work with community-based organizations, than panchayati raj institutions. When the Hariyali Guidelines gave more space to Panchayati Raj institutions in the implementation of watershed programs in 2002, NGOs opposed it (Shah 2003; WASSAN website, http://www.wassan.org, accessed November 2, 2013). Very often NGOs equate people's participation with their own space in the program. Also, most NGOs in watershed development have shied away from politics, and usually view village-level political dynamics as an impediment to their work, leading to depoliticization of the development process itself.

Assessing Civil Society Action in Watersheds

As noted earlier, civil society action has definitely impacted watershed development. In fact, there has been a very close collaboration between the NGOs and the state in shaping watershed development policies and programs in India. A large number of civil society organizations in watershed programs, primarily NGOs, are consulted and some of their learnings have influenced watershed policies and guidelines. The most recent experience in civil society engagement with policy formulation is through the Parthasarathy Technical Committee and formulation of the Common Guidelines. An NGO representative, Dr Aloysius P. Fernandez of the MYRADA, was a member of the Parthasarathy Technical Committee and Dr. Mihir Shah of Samaj Pragati Sahayog was its Honorary Advisor. Prominent NGOs that the Committee met include Development Support Centre, Seva Mandir, Gram Vikas, Agragamee, MYRADA, People's Science Institute, Professional Assistance for Development Action, Aga Khan Rural Support Programme, Watershed Support Services and Activities Network, NM Sadguru Water and Development Foundation, Samaj Pragati Sahayog, Watershed Organisation Trust and Hind Swaraj Trust. The Forum for Watershed Research and Policy Dialogue (ForWaRD) also organized a consultation with the Technical Committee in Pune and tried to put forward its view point.[31] The Committee also organized a meeting in Delhi of "innovative NGOs" and tried to visit many NGO project sites. In fact, the report has acknowledged the inputs received from such a wide consultation. Samaj Pragati Sahayog, who has long experience in the watershed sector, played a very significant role in the functioning of the Committee.[32]

So by and large, civil society organisations (mainly NGOs) have closely collaborated with the state for watershed development. The state has not only given them space and funds to implement projects, but an increasing role in shaping policy. In fact this collaboration is very often so close and uncritical, that mainstream NGOs have started "thinking like the state". There is also a sense of competition among the larger NGOs, to show off how each enjoys a better relationship with the state. By contrast, the engagement of civil society organizations in forestry has been more antagonistic, and the space available to them far more limited.

Inevitably, the question arises as to why watershed development is more amenable to civil society action than, say, forestry. To answer this, we need to re-visit our premise that civil society action is contingent on the bio-physical, social and institutional characteristics and property regimes of the resource concerned; as well as the kind of CSOs involved. In watershed development, the existing property regimes, rights and agency control are rarely seen as problematic. In other words, the watershed development programs work within the existing property regimes, rights and agency control, and do not challenge them. Watershed development is mainly done on private lands and only sometimes on common lands. Benefits from the watershed program basically flow according to the property regime, and the emphasis is on enhancing productivity on private lands. The experience of common property land resource development, barring a few exceptions, has been pretty dismal (Shah 1998). In the case of forests, it has traditionally been more "closed", more controlled by the state through its agency, the forest department. Apart from the existing property regimes and the issue of rights and access, the colonial legacy with regard to forest access and control has also played a crucial role in defining the scope of civil society action.

The issues at stake – which define the boundaries of civil society action as well as the NGO-state relationship – differ in watershed development and forestry. In watershed development, the emphasis has so far been on resource conservation and augmentation. Of course, recently the issue of resource governance in terms of resource use, allocation and access are being included in the watershed discourse, but they have not yet become part of the mainstream debate or practice (Joy et al. 2006). It also follows then that watershed development requires significant funds for resource augmentation and development efforts, and this gives donor agencies greater leeway to have their say. Often, bilateral and multilateral agencies, Indian and foreign donors make the provision of funds conditional on NGO participation in the programs.

Another reason for the uncritical engagement of the NGOs with state in the watershed sector is that many NGOs are dependent on government funds, which brings its own constraints. The government (the Ministry of Agriculture together with the Ministry of Rural Development) is the single largest funder, and accounts for about 70 percent of all funding for watershed programs (Joy and Paranjape 2004).

By contrast, in forestry, the main issue is of recognizing the rights of the local communities (basically *adivasi* communities) over forests. Since colonial times, the state has tried to establish greater control over forests and alienated the local communities. Many civil society organizations are still struggling to ascertain the rights of the local communities over the forest.

The effort here is not to paint a black and white picture or create an either/or situation. For example, following protracted struggle, lobbying and networking, organizations working on forest issues and *adivasis* have been able to force the state to take some of their concerns on board. The Forest Rights Act is an example of this. It is also true that there are organisations working on forest issues who do not have an antagonistic relationship to the state, as they are involved primarily in welfare, developmental or conservation activities. Also, it is not the intention to say that all NGOs in the watershed sector have a cosy, non-problematic and collaborationist relationship with state agencies. The effort here is to take a broad view, study the extent and nature of civil society action in these sectors in relative terms, and then say that the issues on which the civil society action is embedded and the basic characteristics of that sector often determine the scope and nature of civil society action.

Our second premise is that the type of civil society organization involved in watershed development influences the nature of its engagement with the state, as well as the scope and constraints of civil society action. Broadly, development oriented NGOs (going by Garain's typology of CSOs) which are in watershed programs tend to be dependent on state patronage, especially funding. If we plot CSOs working on watershed development issues and forest issues on a continuum of varying degrees of collaboration to varying degrees of confrontation with the state, organized from left to right, then we would find that most of the organizations working on watershed development sector would fall on the left side and most of the organisations working on forest issues would fall on the right side of the continuum. Very few organizations from both the sectors occupy the middle ground, with lower levels of collaboration or conflict. This also partially explains why civil society action could not fully utilize the space available within the system to promote goals like equity, even though watershed development offers a positive sum game.

The development NGOs seem to be filling in the space left by both the state and radical social movements. With the liberalization, privatization and globalization regime, there is a gradual withdrawal of the state from some of its traditional functions, and NGOs (as well as corporate bodies, either through public-private partnerships or privatization of service delivery) are stepping into those spaces. NGOs are increasingly getting into service delivery (Chandoke 2005) and implementation of projects. In fact, NGOs need to re-assess and re-conceptualise their role in watershed development programs. Perhaps they need to change gears, that is, move away from large scale implementation of watershed projects and concentrate more on piloting and grounding innovative ideas and then work as a pressure group to force the state to incorporate these concerns and innovations into the policies.

Though mass-based organizations and new social movements could provide an ideological space for watershed based development, there is a disconnect between these movements and watershed implementation on the ground. These movements have not tried to engage with the nuts and bolts of watershed development. The state, too, does not provide any space for these movements to do so. For example, although the Parthasarathy Technical Committee met a very wide range of NGOs, it did not invite any social movements trying to organize people around drought, equitable water distribution or displacement to the consultative meetings. Neither did any of these movements make an effort to engage with the Parthasarathy Committee and bring on board their concerns. This disconnect needs to be bridged, and unless the social movements show willingness to engage with the substantial issues involved in watershed development, the full potential of the program in terms of the normative concerns of livelihoods, sustainability, equity and participation/democratization may not be adequately addressed. Such an engagement would also slow down the "NGOization" and depoliticization of the development sector.

Finally, NGOs need to critically re-examine their relation to the state. At present, the dividing line in the watershed sector seems to be pretty thin, and there is competition between NGOs to prove who is the closest to the state and who has been able to make the maximum contribution to policy formulation. The state also co-opts the NGOs, and one easy way is to put their members on various committees. The argument here is not against the NGOs relating to state per se, but to consider strategies that enable civil society organizations to be effective, while maintaining relative autonomy.

Acknowledgements

I acknowledge the inputs from my colleagues in the Forum for Watershed Research and Policy Dialogue (ForWaRD) especially Shanachandra Lele, Suhas Paranjape, Abraham Samuel and Amita Shah for their inputs. I have also extensively used the Joy and Paranjape (2004) technical report: Watershed Development Review: Issues and Prospects, Centre for Interdisciplinary Studies in Environment and Development (CISED), Bangalore and various outputs of the FoRWaRD. I also gratefully acknowledge the detailed comments and suggestions of Prof D. N. Dhanagare on the first draft of this article.

Notes

1. I use the forest sector for comparative analysis mainly because both fall in the realm of natural resource management.
2. Meanwhile NWDPRA also revised its Guidelines in 2000 to make the program more "participatory, sustainable and equitable." For details see Government of Indian (WARASA (Watershed Areas' Rainfed Agricultural Systems Approach) – Jan Sahbhagita: Guidelines for National Watershed Development Project for Rainfed Areas (NWDPRA), 2000. The Common Guidelines of 1994 brought five different programs, including the Drought Prone Area Program (DPAP) under one guideline. The Common Guidelines were first revised in 2001, then were radically altered and issued as "Hariyali Guidelines" and became operational from April 2003. The two important Government of India programs – NWDPRA under the Ministry of Agriculture and projects under the Ministry of Rural Development together account for about 70 percent of the funds and area under the watershed program in the country.
3. For a detailed treatment of the theme "watershed development and drylands" see Mihir Shah et al., 1998.

4. Of course compared to the investments in mainstream irrigation sector, the investment in watershed development programs is insignificant.
5. Evaluation studies and reviews like Abraham et al. 2009; Joy and Paranjape 2004; Kerr et al. 1999; Shah 2009 bring out this point.
6. For a detailed discussion on watershed see Paranjape et al. 1998; for watershed classification see Bali, 1979.
7. Land is divided into different classes according to their ability to support crops on a sustainable basis. The most popular classification scheme divides land (soils) into eight classes from Capability Class I through Capability Class VIII in order of decreasing suitability. The classification is based on a joint consideration of texture, soil depth, soil type, slope and erosion characteristics. The eight classes could be grouped in two land use suitability groups, namely, one, land suited for cultivation and other uses (classes I to IV) and two, land not suitable for cultivation but may be suitable for other uses like plantation and so on. For a detailed discussion on land capability classification and related issues see Paranjape, Joy et al., *Watershed Based Development: A Source Book*, 1998.
8. Of course the resource poor sections can derive benefits from the development of common property land resources in a watershed, from non-land based income generation activities promoted as part of watershed development, or through self-help group activities. But these activities do not form the core of watershed development programs.
9. Shah (2003) clubs all these inequalities under the umbrella term "historically disadvantaged." For a detailed discussion see Shah 2003.
10. For a detailed discussion on the evolution of watershed programs since the early 1930s, refer to Shah 1998.
11. This is not to say that there are no problems with the Guidelines. In fact there has been a fair amount of criticism about the Revised Common Guidelines (2001) which further got revised (known as Hariyali) and became applicable from April 1, 2003. In the Hariyali guidelines, though the alleged aim is "to further simplify procedures and involve the Panchayat Raj Institutions (PRIs) more meaningfully in planning, implementation and management of economic development activities in rural areas", the main criticism has been that there has not been enough devolution of powers and also that the space for NGOS, CBOs, etc. has been reduced. For a detailed discussion refer to Anil Shah, "Fading Shine of Golden Decade: The Establishment Strikes Back", 2003 and also the WASSAN website, www.wassan.org for related material on Hariyali workshop reports, recommendations, consultations with CBOs/PRIs/NGOs, concept papers on Hariyali.
12. This is also reflected in the detailed "success criteria" given in the revised NWDPRA guidelines. For details see GoI 2000.
13. For a detailed discussion on normative concerns, see K. J. Joy and Suhas Paranjape, "Watershed development review: Issues and prospects," 2004.

Almost all watershed guidelines mention livelihoods, sustainability, equity and participation, though what they actually entail would differ. Here my effort is to articulate what they should actually mean in the context of watershed development, as a certain minimum expectation.
14. Since men control the cash income, very often the cash they spend on non-essential items like alcohol do not contribute to livelihood needs.
15. It is important to see who participates, and in a heterogeneous, hierarchical society like ours, democracy within the local communities is also important.
16. The discussion on civil society organizations is mainly drawn from Sangeeta Kamat, *Development Hegemony: NGOs and the State in India*, 2002.
17. In the early 1980s, Rajni Kothari and others connected with the Lokayan group brought forward the concept of "non-party political formations," which brought together new women's movements, non-party mass organizations and funded "voluntary agencies" to contrast them with traditional left parties and their mass organizations.
18. Of course this is in terms of public policy and public investment; in fact groundwater accounts for more than 50 percent of irrigated area in the country and most of it is due to private investment.
19. For a detailed discussion on the Narmada Bachao Andolan and the struggles against the Sardar Sarovar Project see Sangvai 2008, Sangvai 2000; McCully 1996.
20. Of course, *Dams and Development: A New Framework for Decision Making* does not seem to have made much of a difference to the thinking of the World Bank, a major player in the water sector the world over, as within six years of this report the World Bank seems to have come a full circle with the book, *India's Water Economy: Bracing for a Turbulent Future* by John Brisco and R. P. S. Malik (2006) and its love for large dams is very evident in the book.
21. For a detailed discussion on Pani Sangharsh Chalwal and its work see Joy and Kulkarni 2007, Patankar and Phadke 2006.
22. For example the Common Guidelines of 2008 has more than doubled the per ha cost of watershed development – from about ₹6000/ha to about ₹13,000/ha and the overall outlay also has been increased significantly (GoI 2008). Of course this is nowhere close to the expenditure on centralized, large projects.
23. The rapid assessments of watersheds in Madhya Pradesh and Maharashtra by ForWaRD bring this out very clearly. The reports are available at http://www.forward.org.in. (accessed November 2, 2013).
24. Some of these are discussed in detail in Joy and Paranjape 2004 and the reviews done as part of ForWaRD work. They are also available at htpp://www.forward.org.in. (accessed November 2, 2013).

25. For a detailed discussion on how processed bio-mass can be used in the various types of construction sectors like buildings, roads and water, see Datye, 1997 and Paranjape and Joy 1995.
26. Of course this is not to say that restructuring (or transforming) the existing property relations are not important. In fact it is important in its own right and has to be pursued accordingly.
27. This is mainly because watershed development is primarily a land based program and only those who own land would benefit in a major way. Coupled with this, the bio-physical characteristics of watershed also result in people who have land at the valley portion of the watershed benefiting much more than those who have land in the upper portion of the watershed, especially in terms of access to water.
28. The NWDPRA guidelines of 2000 are called Watershed Areas' Rainfed Agricultural Systems Approach (WARASA) – JAN SAHBHAGITA Guidelines.
29. Forum for Watershed Research and Policy Dialogue (ForWaRD) is a collaborative initiative of three institutions, namely, Society for Promoting Participative Ecosystem Management (SOPPECOM), Pune, Gujarat Institute of Development Research (GIDR), Ahmedabad and the erstwhile Centre for Interdisciplinary Studies in Environment and Development (CISED) which is now part of Ashoka Trust for Research in Ecology and the Environment (ATREE), Bangalore.
30. ForWaRD organized a consultative meeting with the Parthasarathy Committee on July 25, 2005 at Pune and the paper presented by ForWaRD during this meeting has been published as an Occasional Paper by ForWaRD (Joy et al. 2006). This paper and other research outputs of ForWaRD are available at http://www.forward.org.in
31. The paper presented by ForWaRD in this consultation has been subsequently published as an Occasional Paper (See Joy et al. 2006).
32. For details of the NGO consultation and the type of inputs received from them see GoI 2006. There was a debate about the report in EPW by Shah, Joshi, Ambasta; Joy et al., and Vaidyanathan 2006; also see Reddy 2006 and Shah 2006.

References

Abraham, Samuel, K. J. Joy, Suhas Paranjape, Eshwer Kale, Raju Adagale, and Ravi Pomane. 2009. "Watershed Development in Maharashtra: A Large Scale Rapid Assessment", Forum for Watershed Research and Policy Dialogue, SOPPECOM, Pune. Available online at http://www.forward.org.in (accessed on November 10, 2013).

Ambasta, Ashesh. 2006. "Need to Flesh Out Proposals," *Economic and Political Weekly*, 41(27&28): 2991–94.

Bali, Y. P. (ed.). 1979. *Watershed Management: Concept and Strategy*. Dehradun: Central Soil and Water Conservation Research and Training Institute.

Brisco, John and R. P. S. Malik. 2006. *India's Water Economy: Bracing for a Turbulent Future*. New Delhi: Oxford University Press.

Chandoke, N. 2005. "Seeing the State in India," *Economic and Political Weekly*, 40(11): 1033–39.

Datye, K. R. 1997. *Banking on Biomass: A New Strategy for Sustainable Prosperity Based on Renewable Energy and Dispersed Industrialisation*. Ahmedabad: Centre for Environment Education.

Garain, Swapan. 1994. "Government-NGO Interface in India: An Overview," *Indian Journal of Social Work*, 55(3): 337–46.

Government of India (GoI). 2008. "Common Guidelines for Watershed Development Projects", National Rainfed Area Authority, Ministry of Agriculture, Government of India, New Delhi.

———. 2006. "From Hariyali to Neeranchal: Report of the Technical Committee on Watershed Programmes in India", Department of Land Resources, Ministry of Rural Development, Government of India, New Delhi: GoI.

———. 2000. "WARASA (Watershed Areas' Rainfed Agricultural Systems Approach) – Jan Sahbhagita: Guidelines for National Watershed Development Project for Rainfed Areas (NWDPRA)," Rainfed Farming Systems Division, Department of Agriculture and Cooperation, Ministry of Agriculture, Government of India. New Delhi: Government of India.

Joshi, Deep. 2006. "Broadening the Scope of Watershed Development," *Economic and Political Weekly*, 41(27&28): 2987–91.

Joy K. J., Amita Shah, Suhas Paranjape, Shrinivas Badiger, and Sharachchandra Lele. 2006. "Reorienting the Watershed Development Programme in India", Forum for Watershed Research and Policy Dialogue, SOPPECOM, Pune. Available online at http://www.forward.org.in (accessed on November 10, 2013).

———. 2006. "Issues in Restructuring," *Economic and Political Weekly*, 41(27&28): 2994–96.

Joy K. J. and Seema Kulkarni. 2007."Engaging with the changing water policy discourse in Maharashtra," Pune: National Center for Advocacy Stuides.

Joy K. J. and Suhas Paranjape. 2004. "Watershed development review: Issues and prospects". Bangalore: Center for Interdisciplinary Studies Environment Development.

Kamat, Sangeeta. 2002. *Development Hegemony: NGOs and the State in India*. New Delhi: Oxford University Press.

Kerr, John, Ganesh Pangare, Vasudha Lokur Pangare, P. J. George, and Shashi Kolavalli. 1999. "The impact of watershed development: Results, from a major field survey," in John Farrington, Cathryn Turton and A. J. James (eds), *Participatory Watershed Development: Challenges for the Twenty-First Century*. New Delhi: Oxford University Press.

National Institute of Agricultural Entension Management (MANAGE). 2000. "Common Principles for Watershed Development," Booklet on Common Approach for Watershed Development, Rainfed Farming Division, Department of Agriculture & Cooperation, Ministry of Agriculture, Government of India, New Delhi.

McCully, Patrick. 1996. *Silenced Rivers: The Ecology and Politics of Large Dams*. Hyderabad: Orient Longman.

Omvedt, Gail. 1993. *Reinventing Revolution: New Social Movements and the Socialist Tradition in India*, An East Gate Book. London: M. E. Sharpe.

Paranjape, Suhas and K. J. Joy. 1995. *Sustainable Technology: Making the Sardar Sarovar Project Viable – A Comprehensive Proposal to Modify the Project for Greater Equity and Ecological Sustainability*. Ahmedabad: Centre for Environment Education.

Paranjape, Suhas, K. J. Joy, Terry Machado, Ajaykumar Varma and S. Swaminathan. 1998. *Watershed-Based Development: A Source Book*. New Delhi: Bharat Gyan Vigyan Samiti.

Patankar, Bharat and Anant Phadke. 2006. "Asserting the Rights of the Toiling Peasantry for Water Use: The Movement of the Dam Oustees and the Drought-Affected Toilers in South Maharashtra," in Peter P. Mollinga, Ajaya Dixit and Kusum, *Integrated Water Resources Management: Global Theory, Emerging Practice and Local Needs*. New Delhi: Sage.

Reddy, Ratna, V. 2006. "'Getting the Implementation Right': Can the Proposed Watershed Guidelines Help?, *Economic and Political Weekly*, 41(40): 4292–95.

Rothman, Jack, John Erlich and John Tropman. 1987. *Strategies of Community Intervention, Strategies for Community Organisation*. Micro Practice, Michigan: F. E. Peacock Publishers.

Sangvai, Sanjay. 2008. "People's Struggles in the Narmada Valley: Quest for Just and Sustainable Development", in K. J. Joy, Suhas Paranjape, Biksham Grijja, Vinod Goud, and Shruti Vispute (eds). *Water Conflicts in India: A Million Revolts in the Making*. New Delhi: Routledge.

———. 2000. *The River and Life: People's Struggle in the Narmada Valley*. Mumbai and Kolkata: Earthcare Books.

Shah, Amita. 2009. "Revisiting Watershed Development in Madhya Pradesh: Evidence from a Large Survey", Forum for Watershed Research and Policy Dialogue, Gujarat Institute of Development Research (GIDR), Ahmedabad. Available online at http://www.forward.org.in (accessed on November 10, 2013).

Shah, Mihir. 2006a. "Towards Reforms", *Economic and Political Weekly*, 41(27&28): 2981–84.

Shah, Mihir. 2006b. "Reforms for Watersheds", *Economic and Political Weekly*, December 9-15, 41(49).

Shah, A. C. 2003. "Fading Shine of Golden Decade: The Establishment Strikes Back", in Proceedings of the New Developmental Paradigm and Challenges for Western and Central Regional States in India, National Seminar at Gujarat Institute of Development Research (GIDR), Gota, Ahmedabad.

Shah, Mihir, Debashis Banerji, P. S. Vijayshankar and Pramathesh Ambasta. 1998. *India's Drylands: Tribal Societies and Development through Environmental Regeneration.* New Delhi: Oxford University Press.

Shah, Amita. 1998. "Watershed Development Programmes in India – Emerging Issues for Environment-development Perspectives," *Economic and Political Weekly*, 33(26): A66–A80.

Vaidyanathan, A. 2006. "Restructuring Watershed Development Programmes," *Economic and Political Weekly*, 41(27&28): 2984–87.

World Commission on Dams. 2000. *Dams and Development: A New Framework for Decision Making.* London & Sterling: Earthscan.

3

How an NGO Works the State and Why it Succeeds

A Case Study from Central India

Vasudha Chhotray

Overview

The central proposition that I put forward in this chapter is that even though non-governmental organizations (NGOs) are regarded as the most visible spokespersons for civil society in recent development discourse, they, in fact fundamentally constitute the state. The term civil society has gained prominence in development discourse, as it has come to signify domestic public space where powers "illegitimately" usurped by states can be reclaimed (Khilnani, 2001: 12; Jenkins, 2001). These ideas have been as influential in India as in other nations, where since the 1980s, the proliferation of an "NGO universe" comprising very diverse and indeed dissimilar organizations has been euphorically noted and lauded (Sen 1999). Ample recognition has followed even within the official apparatus of the Indian state, which has duly initiated various programs of state-led reform aiming to introduce decentralized and participatory forms of governance to ward off criticisms against centralized and top-down government. Eager to share in the salutary effects of "civil society," the key component of such initiatives is often the involvement of NGOs of various hues in different capacities. The point simply is the use of the NGO as a metaphor for civil society by officials as much as non-officials.

My purpose in this chapter is neither to present a tirade against the equation of NGOs with civil society, nor to argue against the projection of civil society as a homogenous terrain without any internal hierarchies of power amongst its various constituents. A number of scholars have done this most satisfactorily (see Bebbington and Hickey 2006, Hadiz 2004, Igoe 2003, for example). Moreover, the point has also been made that identifying NGOs as symbolic civil society actors who can achieve objectives like democratization and greater state accountability, follows a rather simplistic dichotomy between the state and civil society, one that is essentially unsuitable for understanding the profound interrelationships between the two in the post-colonial developing world (see Kaviraj and Khilnani 2001). So, for instance, Bayart (1986) and Kaviraj (2001) have described how, following peculiar historical conditions, civil societies in sub-Saharan Africa and India are neither substantive entities, nor armed with limited powers against their respective states. Chatterjee (2004) further rejects the wider applicability of the idea of civil society in India, on the ground that it is inhabited solely by a small section of elites. A broader notion of "political society," Chatterjee suggests, is suitable for understanding how the vast majority of people interact with state agencies to negotiate their claims.

In this chapter therefore, I will not restrict my analysis to the prospects of NGOs in furthering or deepening civil society involvement in traditionally state-led activities. Instead, I will pose a somewhat different set of questions: how do NGOs see the state, their own position within it and most importantly, what do they make of the idea of the state in negotiating a space for themselves to undertake socially transformative development. I think these questions are necessary if we are to understand the character of Indian civil society and the opportunities and constraints available to NGOs, given their often unique position in the interface between governments at different levels (both elected representatives and bureaucrats), local communities and foreign donors. My inspiration for these questions comes from the experiences of an organization in the tribal lowlands of Bagli in Madhya Pradesh that I observed during a period of doctoral field work in 2000 (and have occasionally remained in touch with since[1]). The purpose of this chapter is to analytically explore questions posed through the viewpoint of this NGO's strategies, all along emphasizing the significance of the particular circumstances influencing its experiences. I also hope to more generally justify in the end my choice of questions, and to conclude that the key to the power that NGOs can wield, may depend on the manner and extent to which they are able to "work" the state.

The Context for State-NGO Relationships in India

NGOs have frequently been depicted in a unified manner in development literature. The point that this causes "ambiguity and conflict" about the role of the grassroots sector, has been made forcefully (Kamat 2002, Bebbington et al. 1993). A brief history of non-governmental action in India reveals that a wide range of voluntary initiatives have existed since Independence, and even before it as well. It shows that NGO is a catch-all term that does not adequately capture differences among these initiatives.

The late colonial period witnessed the emergence of several indigenous voluntary organizations for socio-religious reform (Sen 1999). Voluntary organizations espousing Gandhian ideals of local and voluntary action for development were involved in a variety of works. In fact, voluntary organizations found explicit mention in the First Five-Year Plan. Moreover, the Community Development Programme in 1952 was the first instance of an official proposal to involve local people in development. However, these initiatives were less influential, following the formulation of centralized development strategies after Independence.

The problems of centralized development planning were evident in the following decades. They unleashed multi-faceted processes of exclusion, leading to popular mobilization movements by peasants, tribals, lower castes, women and environmental activists (see Ray and Katzenstein 2005; Raj and Chowdhury 1998 and Rao 1978 for good overviews). Peasant mobilizations linked with the left political parties in the country illustrate well the linkages, as also tensions, between electoral and extra-electoral politics. The Communist Party of India led the peasant insurrection in

Telangana (then a part of the princely state of Hyderabad) during 1946–51. Subsequently, differences regarding ideological sympathies with China as a major communist power led to the birth of the CPI (M) in 1964. But the newly formed party was torn apart by disagreements regarding the readiness for "revolution," and, while the leadership of the CPI (M) opted for participating in legislative politics, a host of "extremist" groups within the party remained extremely critical of this move. The Naxalbari peasant uprising of 1967 further exacerbated schisms within the party, and several extremist groups were "either expelled or they dissociated themselves with the Party" (Mukherjee 1978: 24). The subsequent grouping of a number of different extremist groups spearheaded by the Naxalbari and Peasant Struggle Assistance Committee (NKSSS) led to the formation of the All India Coordination Committee of Communist Revolutionaries and the CPI (Marxist-Leninist) in 1969. Naxalites around the country crystallized their support through this new party and stepped up their political momentum by launching radical guerrilla tactics aimed at class annihilation (contrasted with the land eviction focus of their previous strategies). By going "underground", the political movement escalated and spread quickly and by April 1969, there were 46 places in India where the movement had struck, with West Bengal and Andhra Pradesh being its hotspots (ibid.). State response to these developments was brutal. Indira Gandhi left no stone unturned in pursuing acts of repression and torture and by 1973, nearly 40,000 activists were in jails. However, as Mukherji (1978) writes, these did not at all signal the dissolution of the CPI (ML). On the contrary, a very large number of groups emerged, claiming to be the real CPI (ML) and there was a re-emergence of Naxalism both in West Bengal (Nadia) and Bihar (Sahar) in the late 1970s.

Thrnl is a also a rich legacy of non-violent political activism originating in Gandhian thought, as manifest most clearly in the Sarvodaya movement. This movement reached its peak in the 1970s through the Bhoodan or voluntary land redistribution movement initiated by Vinoba Bhave in 1951, and was later taken forward by J.P. Narayan and others. The movement embodied key tenets of Gandhian socialism including the dignity of labor, equitable distribution of wealth and communal self-sufficiency.

An interesting parallel has been drawn between the *gramdan* movement (which attempted to replace individual land ownership with communal ownership) and the early Naxalbari movement in that both eventually broadened out from their initial objectives (Mukherjee 1978). While the Naxalbari movement aimed at total class annihilation in order to go beyond the "narrow economic demands" of previous peasant struggles, the *sarvodaya* movement "relapsed into a mystic trust in a change-of-heart that would usher in a new society" (ibid.). Many groups that have descended from the *bhoodan* and *gramdan* networks continue to function locally, although there is no significant presence of any one group on a national scale.

The 1970s in general marked a very large range of protests against the Indian state, then personified by Mrs Gandhi. In 1975, she had taken recourse to a quasi-fascist, totalitarian mode of governance by declaring a national emergency, to bypass the judgement of the Allahabad High Court setting aside her election to the Parliament. There were radical student movements like the Chhatra Yuwa Sangharsha Vahinee and a range of women's movements such as the Maoist movements in Hyderabad (the Progressive Organization of Women) and Maharashtra (Purogami Stree Sangathana and Stree Mukti Sanghatana), (Ray 1999). Women's issues received national legitimacy through a report on the status of women published in 1974 and also by the UN Declaration of 1975 as International Women's Year. However, women's movements, like other resistance movements, had been driven underground in 1975, but these burst into the political scene after the fall of Indira Gandhi's government in 1977. There is a rich literature documenting the trajectories of women's activism in India (see Ray 1999, Basu 1992 in particular).

By the 1970s, many NGOs in India had started receiving foreign funding from international donors. International NGOs like CARE and OXFAM had also started working in India. In 1976, the state under Indira Gandhi passed the Foreign Contribution Regulation Act (FCRA), which required that all foreign money to voluntary organizations be routed through and cleared by the Home Ministry.[2] This Act was perceived as a sign of state control over NGOs receiving foreign funds (Kamat 2002, Fernandes 1986).

Yet, the state did not indiscriminately suppress all NGOs during this period. It continued to support welfare-oriented NGOs and those that did not challenge its development strategies (Kothari 1986, Franda 1983). In fact, the Third Five-Year Plan (1961–65) reiterated the importance of NGOs in the successful implementation of development plans. The coming to power of the Janata Party in 1977 was clearly the turning point in the history of state-NGO relationships in India. The party comprised thousands of Sarvodaya workers who had rallied around J. P. Narayan against the Congress government, as well as older generation Gandhians, who advocated a greater role for NGOs in development. Some factions of this party may have also promoted NGOs for partisan gains, like the Jana Sangh Party (Sen 1999: 339). The Janata government undertook greater initiatives than its predecessor to encourage voluntary work in the "countryside" as part of its rural improvement program. It was at this time that the Government of India instituted CAPART (Council for Promotion of Participatory Action and Rural Technology) to provide technical and financial assistance to voluntary organizations and workers. As a result, a large number of NGOs mushroomed to avail of new opportunities for development. At this time, several voluntary activists, especially Left supporters of the Naxalite movement, the CPI-ML and Maoist groups, were released from imprisonment. While they did not wish to return either to their fragmented movements or to radical party politics, they were motivated to resuming political work in some form (Kamat 2002: 16–17). This led to a steady burgeoning of independent groups, committed to work among the rural and urban poor, and mostly not linked with any formal political party. Many scholars have called such groups social action groups or non-party political formations (Kamat 2002, Sen 1999, Lewis 1991 and Kothari 1986).

By the 1980s, the NGO universe in India had grown substantially, ranging from welfare-oriented NGOs to those formed by international NGOs, social action groups, NGOs formed with government support for development work, development research organizations or think tanks and smaller Community-Based Organizations (CBOs).[3] There are other voluntary initiatives as well, that are typically not described as NGOs. These include issue-based lobby groups such as large farmers' organizations, and mass movements like the Narmada Bachao Andolan (NBA). The NBA in particular is a striking example of a new social movement that comprises development NGOs, grassroots organizations (GROs), activists and intellectuals, all of whom have come together to oppose the Narmada Valley Project.

This brief discussion indicates that a large variety of voluntary initiatives have existed in India since Independence and even earlier. Many of these are clubbed under the term NGO. Attempts have been made to classify NGOs in India using characteristics such as size, composition, origin, source of funding or ideological orientation.[4] Despite these taxonomies, terms like NGO, CBO, GSO or Grassroots Support Organization, MSO or Membership Support Organization and also CSO or Civil Society Organization are used without further qualification by donor agencies, government officials and scholars in India.[5]

The discussion also shows that NGOs continue to share a range of relationships with state institutions and actors at different levels. NGO activity in development, mostly as implementers of development programmes, whether state or foreign funded, has generally been received favorably. State willingness to involve NGOs in development and relief work has grown since Independence. Simultaneously, NGOs that have adopted a politically sensitive or confrontationist stand against state policies, institutions or actors, have risked dissociation from the state's development programs, and occasionally been repressed. The Seventh plan document of the Government of India even defined NGOs as "politically neutral development organizations that would help the government in its rural development programmes" (GOI 1985, as cited in Sen 1999: 342). This cynical view of politics is pervasive in heavily bureaucratized development endeavors, and was immortalized as the "depoliticization" of development by James Ferguson in his classic study in 1990.

It is this stance that can create tensions for NGOs increasingly being brought within the fold of state-led initiatives for decentralized and participatory governance. So, for example, NGOs that work intimately on resource issues dealing with water or land are more than likely to be presented with inequitable and conflict-ridden situations, and they may be required to adopt positions that are less than welcome within regular government-funded development initiatives. I fully recognise that not all NGOs embrace a confrontationist stance for a variety of reasons, including their funding charters or apprehensions regarding the consequences of upsetting power holders. But for those that do, and are more explicitly "political" in the sense of challenging established status quo, are we to conclude that these organizations also cannot work "constructively" with the Indian state? Are we then faced with the over-characterized antipathy between an ever-democratic civil society and an ever-repressive state, one that is massively popular with foreign funders who seek to fund NGOs in support of participatory development or democratization objectives (see Igoe 2003, Jenkins 2001 for excellent critiques)? Indeed, as these authors and many others have rightly pointed out, there are vast differences even to the extent to which different NGOs follow internally democratic norms and principles. In this context, dichotomies between a repressive and "depoliticising" state and a democratic and "repoliticising" civil society (NGOs) are limited in their usefulness to appraise the activities, concerns and strategies of NGOs. I have therefore decided not to use this theoretical dichotomy, and focus instead on the ways in which an NGO that seeks to be socially effective in fulfilling its stated objectives draws upon the resources of the state in a variety of ways.

Viewing the State

This story is set in the picturesque but poor valley of the Narmada, in the Dewas district of Madhya Pradesh. The Narmada transects Dewas into the plateau (*ghaat-upar*) and the valley (*ghaat-neeche*). The two portions could not have been more distinct, in political, social as well as economic terms. The plateau is fertile, irrigated, industrialized and economically prosperous. Moreover, it is dominated by non-tribal upper castes that are influential in the district's politics, as well as political economy. By contrast, in the valley, large tribal pockets comprising the Bhil and Bhilala tribes are interspersed with an exploitative non-tribal majority. The area is marked by decades of resource degradation as well as political marginalization, particularly of the tribal communities resident there. The tribals of *ghaat-neeche* have been locked into an enduring conflict with their non-tribal oppressors, the roots of which can be traced back to the post-independence settlement process when the forest department took over administration of forest areas, thus dispossessing tribals of their lands. While most tribals in Bagli's 100-village belt were compensated with small plots, these lands were largely dry and of poor quality. Poor tribals practise a combination of rainfed agriculture, wage labour and an annual routine of tortuous migration to the plateau areas during the long, dry summer.

The NGO whose story I am about to narrate, was established in Bagli, a tehsil in Dewas's *ghaat-neeche* area, in 1992.[6] It was not the massive organization it is today, with established links with the state and central governments and various donors. It started out as a small group of eight friends who had met at the Jawaharlal Nehru University in New Delhi, and also the Centre for Development Studies in Thiruvananthapuram, Kerala. The members shared middle class backgrounds. They were highly educated and English speaking, while conversant in Hindi the national and main regional language). Nearly all had full-time academic careers before they decided to work more directly in pursuit of their beliefs.

Although the group had no clear strategy or design, its intent was absolutely clear from the beginning. The members of the group wanted to build a "peoples" organization" that would engage in grassroots work and advocacy.[7] The formation of a "critical mass within policy making, so that marginalized tribal areas would get the benefit of increased state intervention and public investment" was central to the stated discourse of the group, and of the organization it eventually formed. From this point of view, the choice of Bagli was a deliberate one, given its history of political and economic marginalization, acute resource degradation and chronic poverty. The group was therefore keen to engage with the contested and intertwined land, forest and water issues of this tribal area. It soon registered itself as an NGO and with the personal savings of its members, set up a temporary office in Bagli town. The NGO was called Samaj Pragati Sahyog (Support for Social Progress or SPS). Implicit in its name was a discourse of working with the people of Bagli, and it envisaged a very clear role for itself in this respect as the incidents of the following years suggest.

From the outset, SPS espoused a philosophy of "positive engagement" with the state, its policies, institutions and actors. Especially since it was seeking to work amongst exploited tribals, it perceived itself as different from the pro-tribal, radical and violent Naxalite and Maoist modes of activism that are widespread in parts of Andhra Pradesh, Bihar and Odisha, among others. A prominent member of the group said, "It is much easier being an activist. Activists have no stakes and can afford to raise slogans and use other attention-seeking tactics. When you combine activism with serious developmental work, it is the hardest combination."[8] The group did not therefore position itself *against* the state. It did not perceive the state to be a homogenous entity that could be denounced and defeated. On the other hand, it was clear that if there was to be any lasting redress to the exploitation endured by tribals in Bagli, it could only follow from sustained engagement with the state. But as it discovered soon enough, not only was the state highly differentiated, even the idea of the state was far from cogent, and different actors (both official and non-official) internalized and used what they perceived as the state's authority, in different ways.

Onset of Contradictory Relationships

The state in small-town India, eg in Bagli *tehsil*, is typically symbolized by the *tehsildar*'s office, the police *chowki* and a few other government buildings. Bagli is a small market town and the arrival and announcement of SPS aroused noticeable curiosity, even concern. If SPS was not part of the *sarkar* (government) and it was also not there to implement any health or education project, then why was it there? The local MLA was particularly perturbed. The initial forays made by group members on motorcycles to the tribal villages did not help assuage these doubts. The group was looking for a beginning and Neelpura seemed to be an ideal choice.[9]

Neelpura is a small village that lies near the main road and is relatively accessible. It comprises mainly Bhilala tribals and is uniformly poor. Most persons here own between 1–3 acres of poor quality and dry land. A handful of farmers own more than six acres and only three out of the 100 odd households in the village are presently landless.[10] It is socially relatively homogenous in social and economic terms. However, it is under the Bhimpura panchayat, which, a prosperous village dominated by non-tribals. When SPS first tried to make acquaintance with the people of Neelpura, it received a lukewarm response. The respectable Patel family were reclusive, and so was Mahbub Khan, a Muslim and the largest farmer in the village. It received its first anchor in Lakhan Singh, a landless though politically connected tribal man, who at that time was also the sarpanch of the Bhimpura-Neelpura panchayat.

Early hesitation on part of the residents of Neelpura towards SPS soon gave way to eager interactions, especially when the latter offered to dig wells on the private lands of people. The SPS received funds for this purpose from various central government schemes given that water conservation in dryland areas is a significant development objective of the state.[11] By 1995, SPS had established itself more firmly in Neelpura and became known in the Bagli region for its development activities. But few had expected that SPS would raise questions that had been unasked for long, questions that would destabilize the established relationships of power in *ghaat-neeche*, especially since it seemed to be a "regular" NGO implementing central government schemes.

During its implementation of the well digging and water conservation projects in Neelpura, SPS stumbled upon two types of exploitative practices in the region. These revealed the nexus of domination by anti-tribal forces in the *ghaat-neeche* area. It detected that the overall wage structure, especially for public works, in this tribal belt was not in keeping with the equal minimum wage laws of the country enacted in 1948. The large farmers and panchayat sarpanches (acting through contractors), who engaged labourers for the execution of construction works, perpetuated this injustice. The SPS also discovered that land records of poor tribals throughout the *ghaat-neeche* had not been updated in accordance with the Madhya Pradesh Land Revenue Code of 1950, and essential information, such as correct rates for land transactions, was being kept out of their bounds by the local revenue bureaucracy.

This included both the village *patwari* as well as the sub-district magistrate of the revenue division, who stood to gain monetarily from such malpractices.[12] Emboldened by the absence of challenge, these junior state officials had also acquired near autocratic status locally.

In an unprecedented move, SPS insisted on paying equal minimum wages to all its hired labourers, making enemies of large farmers, sarpanches and contractors in the *ghaat-neeche,* all of whom of course had benefited from existing local wage relations. Mahbub Khan and Lakhan Singh distanced themselves from the NGO immediately. This was the time for dramatic developments, and SPS only intensified the confrontation by contacting the District Collector with a proposal to organize a "land records camp" in order to rectify the appalling records situation. The senior most official of the district lent her support to SPS, and in January 1995, such a camp was organized in Neelpura village. It was a huge success, with more than 13,000 tribals traveling far to attend, and the District Collectorate backed it with two additional camps. And so unfolded two entirely contradictory relationships between the SPS and the state actors – on the one hand, junior officials like the sub-divisional magistrate who suffered humiliation and material loss by the organization of the land records camp were drawn into an intensely antagonistic relationship, but on the other hand, SPS seemed to embark on a very cordial and friendly equation with top district and state-level officials, together strategising various redemptive steps in favor of the tribals.

How did the SPS manage to evoke such a reaction from the state in Dewas to rectify old and contentious practices that seemed to be deeply embedded within the prevailing status quo? While a certain ease in communication owing to similar social and educational backgrounds as of the senior state officials in the district may have helped, there were other more important reasons. SPS successfully emphasized the idea of the Indian state as a guarantor of rights. Senior civil servants would have found it extremely difficult to remain inactive in the face of blatant aberrations of existing legislation (in this case the Minimum Wages Act of 1948 and the Madhya Pradesh Land Revenue Code of 1950). The location of the struggle in *ghaat-neeche* and not *ghaat-upar* made some difference to the ability of the district administration to respond favorably to the demands made by the SPS. Greater political stakes embedded in the long history of non-tribal and upper caste domination would have meant lesser space for tolerance of opposition, a point admitted to by both the SPS and district government officials whom I met.

It is true nevertheless that even in *ghaat-neeche* SPS faced tremendous opposition. A group of sarpanches, mobilized by the local MLA, petitioned the then Chief Minister Digvijay Singh, to have the NGO "removed" from the area. Their grounds for this were that the NGO "was corrupt, it bypassed panchayats and misappropriated their money." The local press greatly dramatized these developments through sensational articles, but the SPS also used the media to build a favourable image of itself as a transparent organization that worked in popular interest. The situation was resolved when an enquiry committee appointed by the Chief Minister and headed by the district level panchayat cleared SPS of all charges and publicly commended it for its "good work" which was tremendously helpful. Public ratification by the highest elected body in the district at a time when the state government was vigorosly pursuing decentralization to panchayats in effect placed SPS on the "side" of the state and not against it. With a firm base in Neelpura, whose people rallied behind SPS, and an ascending hold on key officials, both elected and non-elected, made the NGO stronger than ever before.

Working for the State

Success breeds itself, and this could scarcely have been truer than it was for SPS in the summer of 1995. Not long after the organization had been publicly vindicated despite taking a politically contentious stand against the non-tribal exploitative coalition, it received a formal invitation from the Dewas district administration to become a project implementing agency for watershed development projects, under the central Ministry for Rural Development funded program in the *ghaat-neeche* villages. Watershed projects under the Ministry's "Common Guidelines," 1994 (later revised in 2001, 2003 and 2008) are implemented on a micro-watershed basis and aim to treat land, water and forest area included in the watershed on a holistic basis. SPS's selection as a project implementing agency for a state funded and managed development programme was appropriate, given that it was already involved in water conservation in the region. It was particularly significant as well, since it showed that the NGO's resistance to certain types of state practices did not preclude its appointment as a formal agent of a premier state development program. It showed that there were no definite boundaries between NGOs that implement development projects using government money and those that resist specific state practices. It also established that the district administration was quite willing to further its alliance with an organization that had taken on some of the most powerful political forces in the *ghaat-neeche* region, at a time when the interest of politicians and political representatives in newly launched watershed projects was strong. The district administration was under considerable pressure from politicians to "award" watershed projects to electorally significant villages especially in the *ghaat-upar* region. And yet, Neelpura was among the first batch of villages selected for watershed development in Dewas and SPS was its project implementing agency.

SPS welcomed the opportunity. It was not entirely a novel matter for the NGO to work on government sponsored schemes and projects but the involvement with a prestigious, well-financed and well-publicized program was certainly special. The central government's scheme was being implemented under the auspices of a specially created technical mission in Madhya Pradesh. The Rajiv Gandhi Mission had been created on the Chief Minister's directions to allow access to specially handpicked "reform-minded" bureaucrats without being encumbered by regular bureaucratic hierarchies. As a project implementing agency, SPS became associated with the mission. But here too, unlike many other project implementing agencies that tend to be bogged down by the literal requirements of the centrally imposed policy guidelines, SPS adopted a bold interpretation of the latter. It went about implementing the watershed project in Neelpura without caring too much for the detailed instructions prescribed for the constitution of a local watershed committee (i.e., through "consensual" methods in the presence of a public village gathering). As it had already established itself in the village and was acquainted with people here, many of whom were already engaged on a number of other projects, it decided to facilitate the formation of a committee led by individuals who it perceived would be "articulate, command social respect and possess potential leadership qualities" in order to be "effective" and "take contentious decisions." Both the persons selected (as committee secretary and chairperson) belonged to the Patel family, and SPS justified their selection on the grounds that they were widely respected and competent. It also trained committee members to conduct works, pay wages and keep accounts and did away with the (tiresome) requirements of keeping minutes for committee meetings. It believed that "demystification" of the project's objectives would be the best way of ensuring "genuine devolution."[13]

The approach of the SPS as a project implementing agency towards the formation of the Neelpura watershed committee offers subtle but significant insights into its own envisaged position in relation to village residents. It is difficult to ascribe a uniform viewpoint to all its members, and I observed that there were differences in the extent to which they talked about this issue. The project officer, a member of SPS, was relatively forthright in his views that it was "nearly impossible" for them to "create a watershed committee that has an identity separate from the *sanstha*" (organization; also the term by which SPS is popularly referred). Besides, this is also not something the NGO, in its position as project implementing agency was necessarily trying to avoid, following its more generally expressed belief that it was important to create a "cadre of local leaders" who would be "from poorer sections", "committed to village development" and ultimately work through panchayats. The work on watershed projects that these selected persons performed (including different aspects of management and financial accounting) was seen to contribute to their training for panchayat roles (in later years, SPS seriously pursued a range of activities to empower panchayats, more details follow later). However, for many in Neelpura – especially those who had reason to feel left out or disgruntled – the close and cosy relationship between some individuals and the *sanstha* was a very sore matter indeed. Some ventured their own explanations to me about the basis of the NGO's strategy, including that it was able to secure village support by recruiting persons from large families. While individual allegations were difficult to substantiate, it was clear that the *sanstha* did not please everybody in Neelpura.[14] A senior member of SPS said, "We are not trying to create a beautiful Neelpura watershed committee that meets and gives beautiful representation to everyone."[15]

The SPS did not appear to place too much premium on orchestrating "democratic" or "participatory" decision-making by the committee through regular meetings or on formalising these staged rituals through minutes, as required by government guidelines. A senior member explained that it was not their belief to try and impose "our structures onto social situations" and they were in favour of letting the decision-making of the watershed coalesce into the "natural flow of village decision-making." "Just imagine a village where people live and work together," he said, "it's not that they need to have a committee.... There are no watershed committee meetings [any more]. It is a far more broad based structure which is what it should be." At the same time, group members emphasized that rigorous mechanisms had been put into place to ensure accountability of funds.[16] As for transparency, SPS insisted that if all payments were "made in the open in front of everyone, nothing can go wrong." All group members shared the feeling that it was important and necessary for the NGO not to intervene in the village beyond the initial phase of setting up the project and training, and often repeated that villagers in Neelpura no longer wanted senior members of SPS to come to their village for the purpose of committee meetings. It remains however that the entire watershed project was closely identified with its project implementing agency, SPS, particularly following its growing scale of operations in the Bagli area.

An episode where SPS did intervene heavily though, occurred at the very beginning of the Neelpura watershed project. This further established SPS's emerging strategy of using state laws and policies to rectify existing malpractices that worked against the interests of poor tribals. This is extremely important, for so much attention is placed on advocating change in state laws and policies, that it is often forgotten that existing legal and regulatory frameworks may themselves have radical consequences if only implemented properly. While conducting work for the watershed project in Neelpura, SPS discovered that the use of the village *naala* – its only common water body – had been improperly appropriated by a small group of upstream farmers. A farmer called Mahbub Khan drew waters continuously from the surface itself, through *naardas* (underground channels) electric pump sets and diesel engines. With a few farmers siphoning off waters upstream, those downstream had practically no access to running water or the opportunity to recharge their wells. Village livestock were the worst affected, since the *naala* ran dry after the rainy season. SPS concluded that if works were carried out under this inequitable arrangement, then a rich and powerful minority would undoubtedly corner all the benefits. It resolved not to go ahead with project activities until the arrangement had been overturned. It is clear that SPS was attempting to intervene in a highly contentious area, and one that other project agencies may have disregarded. But as the project officer explained, advising villagers on how "best" to deal with conflict was an integral part of SPS's position as the project implementing agency. The Ministry of Rural Development's policy framework emphasizes common property resources. So, interpreting the powers accorded to it within this policy to the fullest, SPS went ahead and mobilized popular opinion in the village to formulate a collective agreement to regulate the use of *naala* waters. A resolution in Hindi was signed by 139 farmers from Neelpura and some adjacent villages which read as follows:

> It is decided by consent (*sarvasammati*) that nobody would ever draw water from the *naala* using a *naarda*. Those farmers who have wells will also not draw water from the *naala* using motors. Those farmers who do not have wells have agreed to draw water from the *naala* on a limited basis according to rules. After the water in the *naala* stops flowing, nobody would draw water from it, irrespective of whether they have wells or not. This water would be kept for cattle only. *All villagers* agree to this resolution (*italics* added).

Mahbub Khan protested vehemently, but under the weight of collective opinion and the NGO's vigilant stand, had to block the underground channels with cement, along with the other farmers. Those who had water in their wells or lands on which wells could be dug had to remove motors from the *naala*. The SPS even constructed additional wells wherever necessary, free of any contributions from the farmer. The *naala* agreement was a matter of tremendous pride for SPS, and it mediated this to the last detail. In the initial days after the agreement, enthused villagers set up a system of rotation to watch the *naala* against possible violators at night. The SPS claims that the agreement benefited everyone although those with lands upstream were at a greater advantage than the rest. Khan protested violently, even taking SPS to court claiming that he had "easementary right" to the *naala* under the Indian Easements Act of 1882. The SPS was well placed to undertake research into the legal history of the *naala*, and was soon able to trounce Khan's claims. The Act allows a single user or group of users exclusive or predominant use over a village resource, on the basis of "long use or prescription," on the grounds that this use has been peaceable, open and uninterrupted for at least 30 years, as an easement and over a resource that is not owned by anyone in particular. But SPS demonstrated that the *naala* was actually owned by the government, which in 1993 had issued an order prohibiting villagers from its use, and Khan himself had claimed right of use only since the last 17 years. Khan was reprimanded for coming to court with "unclean hands" and his appeal for "easementary right" was struck down. This had the effect of upholding the *naala* agreement and effectively altered the local field of power.

The agreement, along with the verdict of the court was a huge blow to Mahbub Khan, both symbolically and materially. Lakhan Singh and other sarpanches in the area, seething from their own separate issues of discontent with SPS, supported Mahbub Khan and sporadic outbursts of anger against SPS continued. After a few such incidents, SPS was enraged and organized a non-violent *dharna* or "sit-in" at the entrance to Neelpura, protesting against the hostility shown by sarpanches towards its members. This *nijbal satyagraha* (non-violent resistance) lasted four days. Sufficiently stirred after the dynamic events of the previous month, villagers from Neelpura expressed their solidarity with SPS. The incident received much attention locally, and concluded on the pleas of several *zilla* and *janpad* panchayat members, who came to appease senior members of the SPS. A month later, in November 1995, the gram sabha of Bhimpura-Neelpura passed the resolution.

With this Act, major local opposition by Mahbub Khan and the sarpanches against the NGO had been crushed. This resolution was important in at least two ways. First, it showed a significant change in SPS's attitude from the time it had raised the minimum wages and land records issue. Eager to avoid allegations that it was bypassing panchayat institutions by assuming a pro-active role in the *naala* episode, SPS was very keen to seek an endorsement of its actions in the panchayat gram sabha. Second, the resolution created a precedent in the region, as never before in Bagli had such a written agreement been attempted in public nor had the institution of the gram sabha been taken seriously enough for such a resolution to be passed at its venue.

The entire experience also had an extremely positive impact on the working culture of inert and defunct panchayat institutions in the area, even beyond the project. In 2000, I was witness to a brewing movement amongst a few tribal *panches* (panchayat members) of Neelpura to challenge the corrupt practices of the non-tribal sarpanch from Bhimpura in a gram sabha, precisely mimicking the bold resolution of the *naala* agreement in a gram sabha five years earlier. Just as SPS had done several times in course of its involvement in Neelpura, they went about collecting signatures in support of their campaign to recall the sarpanch in the middle of his term (recently sanctioned by a radical new legislation in the state). Eventually the expected confrontation in the gram sabha whittled down to a heated argument with the sarpanch who abandoned the gram sabha and left and nothing more came of it. Nevertheless, SPS takes tremendous pride in these developments, attributing it to its insistence on transparent functioning, and the debates on accountability that these had generated.

The *naala* episode reiterated SPS's willingness to take on contentious issues within the framework of a state-sponsored development project.[17] It had in fact acted as a political agency for transforming the local field of power as a pre-requisite to continuing the project. In its emphasis on establishing rules for equitable access to common resources in the watershed, parallels may be drawn between SPS's actions and the initiatives undertaken by Anna Hazare in Ralegaon Siddhi and Vilasrao Salunke in Naigaon through the Pani Panchayat. However, a senior member of SPS unequivocally maintained that they have never been interested in creating a "model village," and that Neelpura presented a very different situation from the much-hyped Ralegaon. He explained that first of all, Anna Hazare was not a "replicable entity" and more importantly, the whole thrust of SPS's work in the region was "towards multiplication" by expanding both watershed work, as well as empowerment of panchayats in a "non-linear fashion."

Consolidating State Support

When the SPS arrived in Bagli in 1992, it was keen to mobilize popular action at the grassroots to campaign for better resources from the state but it did not anticipate stiff resistance from local bodies of the state, the panchayats. It learnt from its early confrontations with the panchayats that these bodies were dominated by entrenched interests and could be an extremely negative force to reckon with. Also, given their constitutional status (and the encouragement they received from the then state government Digvijay Singh's government) they were there to stay. So while it gradually built a popular base for itself, first in Neelpura and then slowly in other neighbouring villages which also benefited from their initiatives, the SPS knew that it could not afford to disregard panchayats in the region and started involving the gram sabha to pass resolutions. It was also encouraged by the demand for accountability in the Bhimpura panchayat, and interpreted this as a positive outcome related to the watershed process in Neelpura. It has since devoted considerable energy towards strengthening panchayats in Bagli. This has served to consolidate the support it receives from both district and state administration.

In 1998, the SPS received a large grant from CAPART which it used to establish a "field station" about a kilometer from Neelpura.

The Baba Amte Centre has been named after a respected social worker who works among tribal people in Madhya Pradesh, and is now a beehive of training activities for panchayats in the region. This important act marked a new phase in the NGO's life history, as it became clear to the people of Bagli and to the *ghaat-neeche* villages that SPS was here to stay. Since then SPS has used the Baba Amte Centre (locally referred to as the *Kendra*) as a base to network with grassroots resources on a national scale. Its practical operations have involved training a large number of development workers in the watershed sector. The agenda for networking grassroots resources included training elected representatives to panchayat bodies. SPS sees itself as an agent for decentralized in the region. It believes that such agents are necessary if the state government's decentralization legislation in favor of local elected bodies is to be effective at all. The district government expressly extended its support to the Baba Amte Centre and local panchayats overwhelmingly expressed their wish to be involved. A new position of strength for SPS with respect to local panchayats resulted in a new toughening of stance against those panchayats that did not extend their cooperation, "we send out a clear message that we want your contribution, only then can we work with you", said a senior member adding, "if you are negative, we are not going to touch you."

The SPS has energetically scaled up its development work, and from a couple of villages in Bagli *tehsil* in the mid-1990s, it now implements a range of development projects in more than 40 villages spread over three tehsils in Dewas and other Khargone districts, with further plans for expansion. Indeed, over the last two decades, it has "grown to be one of India's largest grassroots initiatives for water and livelihood security."[18] Its staff strength exceeds 100 and it has constructed new, impressive offices in Bagli. The main focus of its projects continues to be related to watershed development and drought proofing, and the funding agencies include the state government, CAPART (an old supporter of SPS) and more recently, some foreign donors.

In all these projects, SPS has been emphatic on transparency and has initiated regular public meetings or Jan Sunvaayis ("hearing of the people"). A typical Jan Sunvaayi involves a large public gathering in the village, attended by grassroots workers of SPS and frequently its founding members. They apprise the public of the project's progress and financial status and answer questions from the audience. The SPS hopes that this exercise would promote a culture of accountability among local bodies in the region. This method of accountability is in tune with the idea of "social audit" in the panchayat gram sabha promoted by the state.

For an organization that is so comfortable in speaking a language welcome to state authorities, that is, a positive language of decentralization and the promotion of local institutional capabilities, some incidents indicate that SPS has not shied away from using its voice for issues that are less complementary. The forests of *ghaat-neeche* have long been the site of exploitation of tribals by the State Forest Department and at least two tribal organizations have emerged to promote collective mobilizations to resist such oppression – the Adivasi Morcha Sangathan and Adivasi Shakti Sangathan. In March 2001, the district administration authorized police firings upon tribals in a number of villages in Bagli, with the purpose of evicting them from the land which they had "illegally" occupied. This act was widely denounced in the local and regional media. Accompanying a sense of general outrage was the view (expressed by the "Friends of the River Narmada", a volunteer based solidarity network) that the "attack" by the state had been prompted to counter the growing political influence of the tribal *sangathans*. SPS played an active part in investigating the firings, compiling a detailed report of the atrocities and supporting many tribal families that had been affected.

By lending its support during this very unpleasant, even dangerous confrontation, SPS remained true to its abiding philosophy of working in the interest of the tribals of Bagli. This was necessary from a pragmatic point of view as well. SPS enjoyed widespread support amongst the tribals of *ghaat-neeche* and it was crucial to publicly express solidarity, or risk damaging (even if not losing) the key source of its current strength and legitimacy in the region. In its philosophy of working with the official state apparatus, it positions itself critically as a popular organization, one with mass support, and its interactions with the former depend greatly on the viability of its stated discourse. Its firm stand sent a strong message to the district administration at this time, reiterating clearly that SPS would be unafraid to take sides if the need arose. Even though the key officials in the district administration had changed hands several times since the heady days of 1995, SPS is well known, its history well established, and its key members well respected within the administration. It is difficult to say how things might have fared had the district administration taken a stern view of SPS's stance, especially given its involvement with so many different state funded projects. It remains however that the intensity of collaboration that SPS shares with the state at different levels have accorded to it a very unusual bargaining power.

I have mentioned how over the years, as SPS evolved in its role as an active agency of change within the tribal drylands of the *ghaat-neeche*, it simultaneously involved itself with the creation of lateral networks with other grassroots organizations in the region and in other states as well. But equally, it cultivated important relationships with policy communities around watershed development at the state and national levels, offering technical expertize rooted in its field experiences and experiments in the range of projects that it continues to implement. Watershed development in India has been the subject of fairly intense reform since the early 1990s, reflecting a wider transition from a soil and water conservation program to a comprehensive program for rural development. The SPS has been playing a key role in advising policy reform at the national level, and a prominent part in the 2005 Technical Committee constituted by the Ministry for Rural Development, Government of India, to review all watershed development programmes in the country.

Conclusion

It would seem that the SPS has gone from strength to strength, and is in a formidable position to undertake important local initiatives, as well as negotiate policy change at the national level. It is always tempting, while reviewing the evidence from a case like this, to draw lessons for other organizations. Indeed, this is what I had promised at the beginning of this chapter. I should emphasize, however, that the lessons I draw are not intended to serve as a blueprint for non-governmental action in India. Several aspects of its experience were deeply particular to the nature of its core organization and its operation in a site of subordinate politics where it was able to create spaces for resistance that will not be easy to replicate. Even so, it is possible to take away the essence of its broader approach, especially if it allows us to rethink the terms in which we urge NGOs to assume leadership of "civil society" to influence and reform the state.

So where can we locate SPS in NGO-state relations that I described at the beginning of this chapter? The usual approaches to NGO typologies on the basis of size, strength of membership, source of funds and so on are not particularly revealing beyond providing a factual description of the organization, although there have clearly been important shifts in each of these respects. The SPS has grown considerably in size. Its membership is no longer restricted to its core group, but includes a vast number of "grassroots workers" who are employees of the organization. In addition to state funds, it also receives funds from foreign donors. It is clear that SPS has intensified its involvement with the state in a range of ways, both in Bhopal and Delhi, but what more can we say about the nature of this engagement? Here I argue that it was the particular interpretation of the idea of the Indian state by SPS that played an important part in establishing itself as a key actor in the *ghaat-neeche* region.

Since the very beginning, SPS positioned itself as an integral part of the Indian state. It recognized that although the state was overarching, access to it was anything but uniform and in fact different groups of people enjoyed very differential kinds of access to the state's power and resources. The tribals of *ghat-neeche* had historically been in a subordinate and disadvantaged position in this regard and therefore it saw itself as an agency that would work to alter their situation in material terms. It did not see any virtue in adopting an explicitly confrontationist stand against the state per se, because it acknowledged that the state was an extremely differentiated set of entities, actors and institutions. Some state actors, like the junior revenue bureaucracy in Bagli were venally corrupt and exploitative, and others like the district collector of Dewas (in the early and mid-1990s) were ready to lend their support for effecting positive changes to the condition of poor tribals. It is important to note that the SPS was also not trying to substitute the state with something else since the idea of the Indian state as a guarantor of rights was absolutely central to its own strategy. In taking up contentious issues against local power holders, some of whom were officials of the state at local levels, it was this idea that SPS wielded, compelling more senior officials to lend their support to its demands. Time and again, it was this idea of the Indian state that SPS put to work, in different ways as necessitated by particular circumstances. In its early days, with the minimum wages and land records' incident, it was the *violation* of this idea that SPS complained about, whereas later on, with its initiatives regarding the strengthening of panchayats, it was the *upholding* of this idea that it emphasized. It was this idea of the state that it used to create synergies between a "popular agency" like itself and official state institutions like panchayats, arguing that without these synergies, formal institutional changes like decentralization would not translate into their desired objectives.

Arguably, a close affinity to the idea of the state as the *right* and *proper* guarantor of rights may also serve to circumscribe the kind of issues that an NGO may be able to address and advocate. It may be that certain kinds of difference, marginalization, oppression and exploitation are so embedded within the norms, laws and institutions of the state, that opposition cannot be managed harmoniously or without antagonistic struggles. In these cases, however, the very idea of the state, or indeed the ideas that the state lends backing to, is exclusionary and underpinned by the asymmetries in social power. On the basis of the research I have conducted, I am unable to comment whether or not SPS has consciously or otherwise refrained from engaging with sustained struggles that might have placed it "outside" the reasonable scope of NGO-state interrelationships. I do concede the existence of boundaries within which NGOs in India may find it possible, if not always easy, to challenge wrongful and exploitative practices and even overturn the status quo. I similarly acknowledge that

NGOs which may trespass these boundaries may find themselves outlawed by the very state whose idea they may have previously evoked. As this chapter has tried to show, the boundary itself is contingent on the variable and particular conjunctures of actors operating within dynamic relations of power.

Further, in echoing the idea of the state, did SPS see no difference in the way it operated and staked power from other state officials, whether elected or non-elected? Indeed, SPS harbored no illusions about the basis of its power and sought consciously to embed itself amongst the tribals whom it sought to empower. While it went about obtaining funds from various state and central government schemes for development projects and interacting with important policy makers, it was its large and growing popular base that acted as its principal source of strength. The dramatic events of 1995 have gone a long way in the making of the NGO, and it is this identity that it cherishes and perseveres to sustain. Therefore, even with the physical transformation and escalation of its scale of activities (some would see this transition as a move from "mobilization" to "institutionalization") SPS projects itself as a people's organization and does not want to be seen as a replica of the district administration. A senior member of the NGO stated categorically, "NGOs cannot replace the state." One might add that the usual dilemmas associated with NGO-led development arising from their "withdrawal" from the region of work are not quite relevant here, given that SPS is well and truly ensconced in Bagli. This has undoubtedly had a deep impact on the mobilization of local identities. As we have seen, allegiance to the SPS plays a major part in such processes, and those who have confronted it have found themselves excluded from its range of activities (whether sarpanches or ordinary villagers). Even in 2000, the year of my fieldwork, SPS had a major presence in Bagli and the *ghaat-neeche* area, and this has only grown in subsequent years. Nevertheless, it is well aware that NGOs cannot claim power in quite the same way as official state actors can. Learning from its own experiences of working and surviving in the *ghaat-neeche*, SPS knows that NGOs need to create a space for themselves amidst the prevailing balance of power, and often this process can be exclusionary. It is aware that NGO power is as difficult to sustain as it is to describe, and is cautious about the way it is perceived, taking as much care to forge partnerships at different levels of the state, as it does to build on its popular local standing.

The question may be justifiably asked that in all its attempts to forge a local standing for itself, did SPS prioritize the creation of "self-reliance" among the local people of Bagli's *ghaat-neeche* villages?[19] It is difficult to answer this question in definitive terms, but I will make two observations at this point based on my previous discussion. First of all, as far as articulated discourse is concerned, the different members who comprise SPS always emphasized that while they had a clear role in relation to the local people they worked with (which included spotting and promoting local leaders, offering political entrepreneurship to challenge status quo and training panchayat representatives), they did not believe in intervening in village affairs, especially beyond the initial stage. Their belief that transparency of payments and publicly enacted accountability in watershed projects laterally influence and inspire positive changes in panchayats, also revealed an enthusiasm for local people taking charge and getting involved. Perhaps this combined with their genuine emphasis on local training was the "non-linear" strategy for multiplication of their work that SPS had in mind, as opposed to directly starting up and running more and more projects themselves. But equally, it is hard to deny that the continued and ever strengthening presence of SPS in the *ghaat-neeche* has contributed to its new hegemonic position, making opposition to it somewhat difficult. I am unable to comment further on its new and unfolding relationships with popular actors in the region although that would no doubt be interesting to observe.

SPS has invested considerable energy into its engagement with the state. This engagement has been as much with the state's resources (and most cynical views of NGOs would rest here!), as with its laws, its institutions and most crucially, with its core ideas. Its remarkable fortunes reveal the inextricability of what NGOs can and cannot do, from the constraints and opportunities presented by these very state laws, institutions and actors. Although there may be vast differences in the permutations of circumstances that particular NGOs find themselves in, it may still be possible for

NGOs to work the idea of the state to their advantage. Indeed, if I were to be ironical, the best bet for civil society in India would still have to be the state!

〰️

Acknowledgements

I would like to acknowledge the kind hospitality and support offered by the members of Samaj Pragati Sahyog in Bagli during my field work in 2000. I would like to thank Vijay Shankar, in particular, for his insights and help, and above all for his friendship.

Notes

1. I have not had the opportunity to revisit this NGO since 2000, and would like to clarify that the case study does not fully capture important changes that may have transpired in its practices in the 11 years since, beyond what I have collated through secondary material. While this does constitute a serious limitation in the presentation of this case study, the chapter contains an analysis of this NGO at a formative period in its own history.
2. This followed a public disclosure by the Asia Foundation, a US based agency, that it was funded by the US Central Intelligence Agency which caused public outrage (Sen 1999, p. 338).
3. The records of the Home Ministry showed nearly 12,000 NGOs receiving foreign funding in 1988 (Kamat 2002: 19). There are thousands others that do not depend on foreign money. CAPART itself funds more than 17,000 NGOs for development work (http://capart.nic.in) accessed November 12, 2013).
4. Garain (1994), Robinson et al. (1993).
5. A Membership Support Organization (MSO) is staffed by members who belong to the locality. A Grassroots Support Organization (GSO) may perform functions similar to a MSO, but typically does not comprise the beneficiaries it intends to serve (Carroll 1992). A Civil Society Organization (CSO) is used to denote a whole range of voluntary organizations.
6. This article tells the story of this NGO's interactions with the officials and people of Bagli from the NGO's point of view, given that a central purpose of the paper is to explore how the NGO develops strategies in response to various situations in the course of its work. It therefore does

not fully present the views and voices of the "locals" which also I had an opportunity to observe and document in course of my research. For a fuller exposition of these narratives, refer to Chhotray (2007a).
7. Interview with a group member, Bagli, 2000.
8. Interview, Bagli, 2000.
9. Names of all the villages and respondents have been changed to preserve anonymity to the extent possible.
10. These figures were true for Neemkhera in 2000, and the village's population would have increased since.
11. In 1992, the Tribal Development Department, Government of India, sanctioned INR 1 million for a well digging project. The next year, SPS received a project from the Department of Science and Technology, Government of India, to construct ponds, field bunds, and other water conservation and harvesting structures. In 1993 and 1994, SPS received two projects from CAPART, Government of India, to continue its well digging activities in Neelpura. In this phase of its work, SPS was not resorting to foreign donors for development funds (Interview with group member, SPS, Bagli 2000).
12. They sold land revenue books worth INR 5 at INR 5,000 per copy!
13. Interview with an SPS member, Bagli, 2000.
14. There was a group of discontented individuals, who were united in the fact that they were with the *sanstha* earlier, but later felt terribly bitter and alienated from it. I spoke to at least 10 such persons, and their reasons for discontent vary widely. Some of them are angry about wells not being dug on their lands, others had been distanced from *sanstha* for a variety of reasons, yet others found the *sanstha's* running objectionable, as they found some people in the village were far more favored than most others.
15. The only landless woman in the village was included in the committee, presumably following the guidelines that recommend the reservation of seats for women, but she did not appear to be aware of the committee's work when I interviewed her. The main initiative undertaken by SPS for women in this village was the construction of a *silai kendra* (stitching centre) on the outskirts of the village, where women could go to learn the art, and sell their products on a group basis, sharing the proceeds with the landless woman. However, I heard from the many women I spoke to of the resistance they faced from their husbands on this issue, which led to a conflict and the eventual burning down of the shelter that served as the center! It long served as a contentious issue and that SPS sometimes held its meetings with watershed committee members at the refurbished center irked those who felt excluded from its activities.
16. All records were maintained by a committee member who had been trained to do so, and then audited at the NGO's Bagli office by a senior

SPS member. The two then filled out and submitted the 18 "progress forms" required by the district administration on a monthly basis.
17. When a senior official at the Rajiv Gandhi Mission told a member of the NGO after the dramatic events of 1995, "We had not sent you to do *satyagraha*," the member quipped, "You have not sent us, and if you want your money, please take it back!"
18. http://www.samprag.org (accessed on November 10, 2013).
19. An anonymous reviewer of this article asked, "What would eventually happen to internal vitality, energy, creativity and dynamism of the local people? Are these qualities of the locals affected by SPS's intervention?"

References

Basu, A. 1992. *Two Faces of Protest: Contrasting Modes of Women's Activism in India.* New Delhi: University of California Press and Oxford University Press.

Bayart, J. F. 1986. "Civil Society in Africa," in P. Chabal (ed.), *Political Domination in Africa.* Cambridge: Cambridge University Press.

Bebbington, A. and Farrington, J., Lewis, D.J. and Wellard, K. 1993. *Reluctant Partners? Non-Governmental Organisations, the State and Sustainable Agricultural Development.* London: Routledge.

Bebbington, A. and S. Hickey. 2006. "NGOs and Civil Society," in D. A. Clark (ed.), *Elgar Companion to Development Studies,* pp. 420–22. Cheltenham: Edward Elgar Publishing Limited.

Carroll, T. 1992. *Intermediary NGOs: The Supporting Link in Grassroots Development.* Connecticut, West Hartford: Kumarian Press.

Chatterjee, P. 2004. *The Politics of the Governed: Reflections on Popular Politics in Most of the World.* New York: Columbia University Press.

Chhotray, V. 2007. "Political Entrepreneurs or Development Agents: An NGO's Story of Resistance and Acquiescence," in Bebbington, A., Hickey, S. and Mitlin, D. (eds), *Can NGOs make a difference? The Challenge of Development Alternatives,* pp. 261–68. London: Zed Books.

Fernandes, W. 1986. "The National NGO Convention: Voluntarism, the State and Struggle for Change," *Social Action,* 36(4): 431–41.

Franda, M. 1983. *Voluntary Associations and Local Development: The Janata Phase.* New Delhi: Young Asia Publishers.

Garain, S. 1994. "Government-NGO Interface in India: An Overview," *Indian Journal of Social Work,* 55(3): 337–46.

Hadiz, V. R. 2004. "Decentralisation and Democracy in Indonesia: A Critique of Neo-Institutionalist Perspectives," *Development and Change,* 35(4): 697–718.

Igoe, J. 2003. "Scaling up Civil Society: Donor Money, NGOs and Pastoralist Land Rights Movement in Tanzania," *Development and Change*, 34(5): 863–85.

Jenkins, R. 2001. "Masking 'Governance' for 'Politics': Foreign aid, Democracy and the Construction of Civil Society," in S. Kaviraj and S. Khilnani (eds), *Civil Society: History and Possibilities*. Cambridge: Cambridge University Press.

Kamat, S. 2002. *Development Hegemony: NGOs and the State in India*. New Delhi: Oxford University Press.

Kaviraj, S. 2001. "In Search of Civil Society," in S. Kaviraj and S. Khilnani (eds), *Civil Society: History and Possibilities*. Cambridge: Cambridge University Press.

Kaviraj, S. and S. Khilnani (eds). 2001. *Civil Society: History and Possibilities*. Cambridge: Cambridge University Press.

Khilnani, S. 2001. "The Development of Civil Society," in S. Kaviraj and S. Khilnani (eds), *Civil Society: History and Possibilities*. Cambridge: Cambridge University Press.

Kothari, R. 1986. "NGOs, the State and World Capitalism," *Economic and Political Weekly*, 21(50): 2177–82.

Lewis, P. 1991. *Social Action and the Labouring Poor: An Experience*. New Delhi: Vistaar.

Mukherjee, P. N. 1978. "Naxalbari Movement and the Peasant Revolt in North Bengal," in M. S. A. Rao (ed.), *Social Movements in India*. New Delhi: Manohar.

Ray, R. 1999. *Fields of Protest: Women's Movements in India*. Minneapolis and London: University of Minnesota Press.

Raj, S. L. and A. R. Choudhury (eds). 1998. *Contemporary Social Movements in India: Achievements and Hurdles*. New Delhi: Indian Social Institute.

Rao, M. S. A. (ed.) 1978. *Social Movements in India*. New Delhi: Manohar Press.

Ray, R. and M. F. Katzenstein. 2005. *Social movements in India: Poverty, Power and Politics*. Oxford: Rowman and Littlefield Publishers.

Robinson, M. 1993. "Overview: NGO-Government Interaction in Asia," in J. Farrington, D. J. Lewis, S. Satish and A. Miclat-Teves (eds), *Non-Governmental Organisations and the State in Asia: Re-thinking Roles in Sustainable Agricultural Development*. London: Routledge.

Sen, S. 1999. "Some Aspects of State-NGO Relationships in India in the Post-Independence Era," *Development and Change*, 30: 327–55.

PART II

Foreign Assistance and Water Governance

4

Water Provision and Modernity

The Consequence of Foreign Aid in Nepal

Sudhindra Sharma

Overview

Water provision has become a major area of intervention by the state in developing countries. The state, through its various ministries, departments and line agencies or agencies working with the government in an unofficial capacity, is in the business of providing domestic water to its population. But even before the modern state began engaging in water supply, people had been obtaining and managing water. For instance, in Nepal, various sources of water were in use. For drinking water purposes, *dhunge dhara, kuwa* and *inar* were used,[1] for religious purposes *nadi, khola* and *kunda* were used[2] and *pokhari, khare* and *kulo* were used for other household purposes.[3] Pokhari is a pond that has still water collected from excess water during the rainy season while khare is a stream that has water only during the rainy season. Kulo is a man-made temporary irrigation canal. All the three, pokhari, khare and kulo are used for the purpose of irrigation (Campbell 1973). This raises the question that if people in Nepal, as in other developing countries, have access to water, what is the provision of domestic water – *khane pani* – all about?

The official reason for state involvement in water supply and sanitation is to ensure "adequate water supply and sanitation services" and "adequate coverage of such services." Interventions are called for because of limited accessibility to safe and hygienic water; the incidence of child mortality arising from water-borne diseases, inadequate sanitation facilities and the ensuing health hazards; freeing especially female labor spent in acquiring water, for agricultural and domestic work instead, and relieving the drudgery of women who must fetch water over long distances. However, before "inadequate services" and "inadequate coverage of services" became the rationale for intervention, premier agencies in Nepal's water sector were primarily responsible for providing "piped water." In measuring its achievements or charting its future plans, estimates were invariably made regarding the percentage of people who had access to "piped water" and those who did not.

With "safe water" replacing "piped water" as the raison d'être for intervention, concerns about inadequate services and coverage of services have dominated deliberations, actions and institutional practices. The issues pertaining to institutional modalities and resource mobilization strategies for expanding services and coverage of services are discussed here. By examining the history of water supply and sanitation in Nepal, this article argues that state provision of water supply in the hills and plains of Nepal is as much about becoming modern, as it is about saving labor, reducing drudgery and increasing convenience. The article begins by discussing literature and ideas from the "anthropology of modernity" and then goes on to examine the history of water supply and sanitation in Nepal before and after 1951. Focusing primarily on the period following 1951, it examines the budgetary allocations for the sector in the five-year plans, the government's economic surveys and unpublished reports and documents. The article concludes with a note on the consequences of water provision by the Nepali state.

Development as Discourse: Concepts and Methodological Issues

Usually the term "discourse" refers to "a formal discussion in speech or writing." But here the term is used in the broader sense that Michel Foucault suggested, including language, practice and institutional regulation, and the relationship between knowledge and power in the representation of social reality.[4] A genre of literature has emerged within sociology and anthropology in the last decade, which focuses not on the people to be "developed," but the apparatus doing the "developing." A few samples – global and in Nepal – are mentioned ahead.

A pioneering work is James Ferguson's *The Anti-Politics Machine: "Development", Depoliticalization and Bureaucratic Power in Lesotho* (1994). The study looks at the deployment of the "development apparatus" in Lesotho, a small land-locked country in southern Africa during 1975–84, how it works and what are its effects. By focusing on a particular rural development project funded jointly by the World Bank and Canadian aid, Ferguson shows how "development institutions" generated their own discourse, which was different from the academic discourse of Lesotho. Ferguson argues that the development apparatus is an "anti-politics machine," depoliticising everything it touches, while politically expanding state power.

Another outstanding work is Arturo Escobar's *Encountering Development: The Making and Unmaking of the Third World* (1995). While Ferguson focused on the discourse of development in a particular place and time, Escobar provides a general view of the historical construction of development in the Third World. Escobar speaks of development as a historically singular experience and analyses it in terms of three axes: the forms of knowledge that refer to it, the system of power that regulates its practice, and the forms of subjectivity fostered by this discourse. The author shows how the ensemble of forms found along these axes constitutes development.

Encountering Development is also anthropological in that it is situated among current attempts to advance anthropology and cultural studies as critical, intellectual and political projects. It provides a new space for a discipline that had until recently only two courses to choose from: that of a "mid-wife," i.e. those who have worked inside development to smooth the transition from tradition to modernity, and that of a "romantic" outside development, championing the authentically indigenous. It looks at the ways in which development operates, with regard to cultural contestation and identity construction. It seeks to contextualize the era of development within overall modernity.

Most sociologists and anthropologists studying development in Nepal have sought to facilitate the development process by trying to better understand the complexity and diversity of the local context. While most of this genre of literature is to help development planners, articles published by Stacy Leigh Pigg (1992, 1993) attempt to step out of the development paradigm altogether and focus instead on how local knowledge and practices are represented in the development discourse. Pigg specifically looks at how the ideology of modernization is assimilated into local culture.

In *Unintended Consequences: The Ideological Impact of Development in Nepal* (1993), Pigg observes that the questions on the role of development in Nepal have been limited by the state rhetoric of development. The terms through which the development establishment justifies itself have defined the debate. The state discourse on development has made possible only two positions – one could be either for development or against it. What is missing in this polemic is a complex, historical understanding of the many levels at which development activities affect a society. Given that development is as much about rationality and ideology as a means to improve the conditions of the poor, the understanding of development as a discursive practice is missing in the debate on development in Nepal, Pigg argues.

Discussions on development in Nepal have largely been in terms of a framework that separates development from the society in which it operates. An analysis of how development activities themselves have shaped Nepali society as a whole, have largely been left out. In order to understand development as a rational practice, Pigg calls for a social analysis of development.

This could include understanding, for instance, what the term *bikas* has come to mean for the locals. True to its Sanskrit roots, bikas in Nepali signifies growth, evolution and development. In everyday parlance, however, bikas is usually a loose term for commodities, benefits or programs, especially those coming from outside. Significantly, it is the non-local origin of these things that makes them bikas. Ironically, even the government is seen as an "outside organization," delivering bikas programs and benefits. This is not due to misinformation, ignorance or lack of political consciousness, argues Pigg, but is rooted in social experience.

Pigg further argues that development language provides new categories in which existing social differences are understood. It fosters an ideological representation of society through an implicit scale of social progress. Bikas reflects a powerful social ideology and according to Pigg, development reinforces social differences in Nepal. The notion that some people are inherently more "developed" than others, echoes Hindu concepts of caste superiority. Markers of class differences also often reflect differences in development.[5]

The issues of translation and local identity-formation that Pigg describes need to be probed more deeply, along with the relationship between bikas and other markers of social difference like ethnicity, caste and class. Following Ferguson's account of the development discourse in Lesotho – how the development discourse constructs Lesotho as an enormously promising candidate for the sort of intervention a "development agency" is capable of launching – the present study looks at how aid discourses construct the domestic water sector in Nepal, as a candidate for its apolitical and technical "development intervention". While I share Ferguson's concerns on discourse and representation, my research includes an exploration of the agents and interests that animate these processes.

Escobar's notions of the "visibility" and "invisibility" of certain discourses, narratives as histories woven from fact and fiction, of deconstructing texts, and his notion of development as a historical construct that can be studied in terms of institutionalization and professionalization, is also useful for this study.

History of Water Supply and Sanitation

Nepal is a highly aid-dependant state, and thus suffers from a severe handicap that a "normal state" does not. A normal state runs on the basis of revenue collected from taxes paid by citizens. It is usually run by elected representatives and tends to be accountable to the people. While paying tax is the responsibility of the people, demanding services from the government is their right.

In an aid-dependant state, the authority and responsibility of the state vis-à-vis citizens are nebulous. The state is run not so much on the basis of taxes from citizens, as aid from donors. Donor agencies have a greater say in state affairs than citizens' voices. Since people's representatives in the government are not able to give priority to their concerns, their credibility weakens, and the loss of political credibility can eventually lead to the de-legitimization of the state itself.

Economic, social and other parameters underscore Nepal's status as a highly aid-dependant state. In 2005 it received a total overseas development assistance of $428 million – 34 percent of central government expenditure or 60 percent of total capital expenditure (excluding recurrent expenditure). The official overseas development assistance it receives is 8.3 percent of its gross national income while per capita aid is $16 – the highest in south Asia. Nepal is on its way to becoming a highly indebted country. Its external debt was $3,354 million in 2005, making per capita debt at $121. Debt servicing accounts for 19 percent of the annual government expenditure (World Bank 2007a).

Nepal is classified as a "least developed country" or "low income country"; its per capita GDP was $270 in 2005. While the economy had grown at 4.9 percent in the 1990s, its growth decelerated to 2.6 percent in the 2000s due to the Maoist conflict that engulfed much of the country. Nepal has a population of 27 million, which is increasing annually at 2.1 percent. The average life expectancy is 62 years for men and 63 years for women. Foreign direct investment is virtually non-existent (World Bank 2007b). The Nepali state's dependence on aid is a structural constraint on the state's capacity to act in general, including the water supply and sanitation sector.

The history of rural water supply and sanitation in Nepal can be divided into five phases where it reached different social strata – (a) the aristocracy, 1880s–1950 (b) urban areas, 1951–1970 (c) district headquarters, 1971–1980 (d) outlying areas, 1981–1990 and (e) further outlying areas, 1991–2002.[6]

Access to piped water in the time of the Ranas was a status symbol, and only a few families with close connections to the prime minister had taps in their homes. When the Ranas were ousted in 1951, it brought together two forces that significantly shaped the country's political history for the next 50 years. The first was a new set of political leaders who had been influenced by the anti-colonial movement in India and aspired to modernize Nepal. The second was the beginning of the international aid era.

Those who fought for political change in 1951 were not only energized by their encounter with the anti-colonial and democratic movement in India, but impressed by the modern amenities introduced by the British in India as well. When Nepal formulated its First Five-Year plan and joined the Colombo Plan (for inter-governmental cooperation for economic and social development among member nations in the Asia-Pacific) in the early 1950s, resources were made available through foreign aid. The US President Harry Truman formally launched the American aid program through his inaugural address in 1949, which outlined the Point Four Program.[7] Investments in domestic water supply and other areas of welfare were possible because of foreign aid. Table 4.1 summarizes the total investments in the various five-year plans and the proportion allocated to domestic water, while the next section outlines the history of the water supply and sanitation sector.

Table 4.1
Allocations for Drinking Water Supply in Various Five-Year Plans in Nepal

(INR in millions)

Five Year Plans (FYP)	Duration	Total Expenditure	Drinking Water	Percentage
First Plan	1956–1961	330	NA	NA
Second Plan	1962–1965	600	24.00	4.00
Third Plan	1965–1970	2500	25.00	1.00
Fourth Plan	1970–1975	3540	37.26	1.05
Fifth Plan	1975–1980	11404	248.00	2.17
Sixth Plan	1980–1985	33940	1,000.00	2.95
Seventh Plan	1985–1990	50410	989.00	1.96
Eighth Plan	1992–1997	170332	6,273.00	3.68
Ninth Plan	1997–2002	372711	11,902.00	3.19

Source: Various Five-Year Plans.[8]

Piped Water to the Aristocracy (1880s–1951)

In the urban centers of the Kathmandu valley, stone water spouts (dhunge dhara) supplied water for domestic use to the majority of inhabitants.[9] The dhunge dhara, still active, supplied water through an intricate underground network of long-distance conduits. In the hills there was no such water supply system and generally spring sources were tapped. In the hills and the Terai, the domestic water sources varied. Kuwa and inar provided drinking water, while nadi, khola and kunda for religious purposes, and pokhari, khahare and kulo provided water for other household needs.

The Rana Prime Minister Bir Shumshere constructed the first modern water supply scheme in Kathmandu valley with the help of a British engineer. The Bir Dhara Works, completed in 1891, tapped the headwaters of the Bishnumati river in the Shivapuri hills, north of the valley to bring piped water to his palace. Another Rana Prime Minister Chandra Shumshere expanded the piped water supply system. Bhim Shumshere built the Tri Bhim Dhara in 1928, and set up a Pani Goswara, a water works office, to manage water services in Kathmandu, Patan and Bhaktapur. Thus, the Rana rulers introduced piped water systems for the benefit of their families and associates in the Kathmandu valley, while the rest of the inhabitants managed with the ancient water spouts.

In the hills and the plains, women walked, carrying water from the sources to their homes; the settlements usually being at a higher level than the sources. If there were suitable water sources located higher than the settlements, these were often quite far off. Piped water supply from higher sources has been the hallmark of the rural domestic water sector since 1951.

Piped Water to Urban Towns (1951–70)

During this period, the main emphasis was supplying piped water to urban centers. Drinking water, along with irrigation, came under the Public Works Department (PWD), Ministry of Agriculture, Transport and Construction. The planning documents of this period speak of "supplying…gallons of water to…towns" and their costs, but do not mention its goals or policies and strategies to achieve them, nor do they mention sanitation and sewage services. A review of these documents does not reveal how the population was managing its total domestic water requirements.

India was one of the donors supporting domestic water supply with the construction of drinking water projects in different parts of the country. In 1954, the Government of India pledged INR 5.7 million for village-level drinking water projects and small irrigation schemes. This pledge was a result of the Kosi Agreement and was meant to assuage Nepali feelings of having received an unfair deal in the agreement,[10] which set out to build a barrage, embankments and a dam across the Kosi, to prevent floods and provide irrigation and power to Nepal and India. Drinking water plants in Sundarijal, Panauti, Bishnumati and Chhatra were subsequently completed through Indian aid.[11]

The Development Board Act was passed in 1956, facilitating the entry of multilateral lending agencies in Nepal, though the actual involvement of agencies such as the World Bank and the Asian Development Bank began only in the early 1970s.

Urban Towns to District Headquarters (1971–80)

Drinking water was included under social services in this period, along with education, health and panchayats. The Department of Water Supply and Sewerage (DWSS) was established in 1972, and the Pani Goswara, established in 1929, was amalgamated into it. The emphasis continued to be providing drinking water services to urban centers, although less urbanized district headquarters also began receiving some attention. As access to clean drinking water was a core focus of the basic needs program, a separate water department was created, and the budget allocation for the water sector increased.

In the early 1970s two key programs provided drinking water services. One was the DWSS, with a mandate to provide drinking water to communities of more than 1,500 people. The other was the UNICEF-supported Community Water Supply (CWS) program implemented by the Ministry of Panchayat and Local Development (MPLD), which was responsible for providing community-based water supply systems to settlements of less than

1,500 people. Thus, while the DWSS provided drinking water services to district headquarters and municipalities, the CWS serviced rural communities. The latter also trained users in building, operating and maintaining water supply systems. The DWSS, on the other hand, distributed work in the form of contracts and did not involve users in the implementation process, other than as petty contractors. The reason for the CWS being in the water supply and sanitation sector was as an entry point for reducing child mortality and improving health.[12] But both programs were independently justified for meeting basic needs.

The World Bank entered the urban water supply sector in 1972, providing the first loan to improve Kathmandu's drinking water and wastewater services. The first phase of the World Bank-funded Water Supply and Sewerage Project began in Kathmandu valley and Pokhara in 1974. The World Bank has concentrated on the urban sector, with one exception, while the Asian Development Bank, which came in later, has focused on the rural sector.[13]

The aims of the drinking water sector were specified in plan documents for the first time. Reasoning that investment in drinking water and sewerage would help meet the basic needs of health and cleanliness, the Fifth Plan document (1970–75) aimed to provide piped drinking water and sewerage systems to a majority of the urban population. From 1976, the Swiss NGO Helvetas supported the CWS in the Western Development Region. From 1981, it was renamed the Community Water Supply and Sanitation Program.

District Headquarters to Outlying Areas (1981–90)

The fourth phase coincided with the beginning of the International Drinking Water Supply and Sanitation Decade. During this period, social services, including drinking water supply, became the third most important sector for investment (24 percent), though in terms of actual outlay, it ranked first (31 percent), surpassing transport and communications and agriculture and irrigation. In this plan, the drinking water sector received an average 3.5 percent of the budget outlay.[14]

Globally, the International Drinking Water Supply and Sanitation Decade target was universal access to drinking water supply by 1990. In Nepal, the plan sought to provide drinking water to 67 percent, or two-thirds of the population by the same year. In order to meet the decade and plan targets, investments in the water sector were substantially increased.

In 1985, a mid-term report stated that the DWSS and the Community Water Supply and Sanitation would not be able to meet the ambitious targets and suggested downscaling the aims. It also set a framework for involving more donors and NGOs to meet the targets, with NGOs serving populations of less than 500.

The major donors in the drinking water sector during this period were the Asian Development Bank (ADB), United Nations Development Programme (UNDP), United Nations Children's Fund (UNICEF) and a few bilateral development agencies. The first loan for rural water supply, the First Rural Water Supply Sector Project (1985–93) was signed between His Majesty's Government of Nepal and the ADB.

This project, implemented by the DWSS , was launched in 22 districts of the Mid- and Far-Western Development Regions. The Second Rural Water Supply Sector Project (1989–95), also signed with the ADB, was launched in 35 districts of the Eastern, Mid- and Far-Western Development Regions of the country. New policy guidelines required the department to adopt a more community-oriented approach, working closely with water and sanitation committee members. The thrust was on constructing smaller, decentralized schemes, and training committee members to administer schemes.

Many NGOs affiliated to the Social Services National Coordination Committee entered the water sector in the 1980s, notably Water Aid and Redd Barna (Save the Children Norway) in 1987; the Associazione Internazionale Disarmo e Sviluppo (DISVI), an Italian volunteer agency; the Lutheran World Service and the Cooperative for American Relief Everywhere (CARE), Nepal, all in 1989. Some agencies like the Nepal Red Cross Society had been engaged in the sector even earlier; other NGOs like Redd Barna, which had been involved in various community development activities earlier, now focused on rural drinking water and sanitation.

In 1987, all rural water supply programs were brought under the Ministry of Housing and Physical Planning, with the DWSS appointed to lead the sector. The Community Water Supply and Sanitation program, supported by UNICEF and Helvetas, was merged into the DWSS. In 1988, the Ministry of Housing and Physical Planning issued a directive changing the role of the DWSS from an implementer to a facilitator.

Further Expansion into Outlying Areas (1991–2000)

The last water phase coincided with the Eighth Plan. Its objective was to (a) provide drinking water to 72 percent of the population, in consonance with the target of providing this service to the entire population in the next 10 years and (b) provide basic knowledge of sanitation and environment protection to the masses. The plan spelt out policies and arrangements for implementing the projects through NGOs, private entrepreneurs and users.

During this period, the ADB gave a loan for the Third Rural Water Supply Sector Project (1992–97), implemented by the DWSS in 40 districts of the eastern, mid- and far-western development regions of the country. The Fourth Rural Water Supply Sector Project (1997–2001), implemented towards the end of the plan period, consolidated the community empowering approaches of the third phase.

The Finnish aid-funded Rural Water Supply and Sanitation Project made its debut as a significant player in the rural domestic water sector during the Eighth Plan. While its preparatory work began in 1989, implementation began in 1990 and by the time the project ended in 1996, it served 2,37,000 people in 54 village development committees in Lumbini Zone, through 110 drinking water and sanitation schemes.[15]

Janatako Afno Khane Pani Ra Sarsafai (People's Own Drinking Water and Sanitation, JAKPAS), a pilot project supported by the World Bank hitherto limited to urban water supply went rural in 1993. Conceived as an autonomous, rural water supply and sanitation fund to support demand-led, community-based water and sanitation activities, JAKPAS worked between 1993–95. The pilot project concentrated on a few hill districts near Kathmandu valley and the institutional arrangements, operational strategies and funding mechanisms found effective were to be implemented in 1996.

The Eighth Plan period witnessed further consolidation of the work of NGOs like Nepal Water for Health (NEWAH), which grew out of Water Aid, Nepal Red Cross Society and new entrants like Action Aid. One of the earliest entrants in the field, UNICEF, began to withdraw, since many agencies were now providing drinking water and sanitation services.

The approach to the drinking water and sanitation sector in Nepal has been largely influenced by the issues dominating international development and environmental agenda. These include the following:

(a) Mar Del Plata Conference, 1977, which emphasized water quality and social aspects of development;
(b) Declaration of the 1980s as the international decade for water supply and sanitation;
(c) Brundtland Commission Report, 1987, which connected the water problem to larger development and environmental concerns;
(d) Dublin Conference on Water and the Environment, 1992, that stressed management of water at the lowest appropriate levels;
(e) Earth Summit, Rio de Janeiro, Agenda 21, 1992, emphasizing freshwater issues, demand driven by users and their willingness to pay, rather than agency targets;
(f) Noordwijk Conference, 1994;
(g) Habitat Agenda Istanbul, 1996, that recommended the government's role as a regulator;
(h) Maracas Conference, 1997, on suitable water policies for the next century.

Water Coverage and Services

The generic problem of rural water supply and sanitation in Nepal has been defined as one of "inadequate coverage and low service level." This encouraged better water and sanitation coverage through pipes and tube wells, and better services. Let us learn about its effectiveness.

As this author has pointed out in the book *Procuring Water: Foreign Aid and Rural Water Supply in Nepal* (2001), this policy conceals as much as it reveals. What it hides is how the intended users of the water and sanitation schemes rationalize the need for external intervention. While water accessibility is the main problem in the hills, water quality is the problem in the Terai. The way in which water projects are perceived by the agencies and users may not be aligned.

As pointed out in the book, even where hill communities had access to spring sources, they were keen to have piped water because the taps would be closer home and more convenient. In addition to more reliable, daily, year-round water supply, it also meant less drudgery carrying water.

In the Terai, many shallow tube-wells and dug-wells meant that domestic water supply was not a problem. The project documents state that the problem is one of water quality. The locals, however, had not asked for external assistance to improve the water quality; rather the services of the NGO were employed to convince the locals of the need for external intervention to improve the allegedly poor quality of water supply and sanitation.

The formulation of the problem as "inadequate coverage and low services" justified the continuation of state structures and aid agencies expanding water supply and sanitation coverage and service levels. The government wanted to meet the basic need of water through a piped network. Water delivery was seen as a hydrological and engineering exercise, headed by civil engineering cadres and the DWSS, with health and hygiene becoming peripheral.

In other words, water supply and sanitation became a procurement problem. While obtaining water through the piped network entailed considerable difficulty, obtaining water meant tangling with the suppliers of pipes, cement, equipment and inevitably, commissions.

The framing of the issue in purely technological terms was advantageous to a hardware-oriented department and over-manned, underpaid bureaucracy, whose main source of surplus income came in the form of commissions from procuring more pipes, cement, rods, hardware equipment and transport services. The department's role in the water sector increasingly came under criticism from other players in the field.

Partly as a response to this criticism, by the end of the 1990s, the DWSS reformulated its role to provide reasonable access to minimum water supply and sanitation services. It envisaged three levels of water supply and sanitation services – excellent, good and minimum – based on indices such as daily amount of water consumption, quality of water, time period of daily supply, accessibility and reliability. The Ninth Plan(1998–2003) envisaged delivering drinking water to all and improved service levels in the next 20 years. The reformulation of its goal as ensuring minimum service standards could also be seen as a departmental strategy to continue being in business. According to the department's own data, water supply coverage, which had reached 61 percent in 1997, was expected to reach 100 percent by 2003; it may have been dissolved, had it not reformulated its role.

Aid agencies and government structures remain in business as long as funds are allocated for a demonstrably worthwhile enterprise. During the 1990s, reducing poverty in developing countries became a priority for many donors. Water supply and sanitation projects quickly justified investment in this sector for their alleged role in reducing poverty.

Conclusion

A global discourse can be discerned in the history of domestic water supply in Nepal. The values, preferences and ways of dealing with water formulated in the West, mediated through foreign aid, work their way through the development apparatus, affecting water policies, and eventually acting upon local values, preferences and usage of water through specific projects. Local culture has already internalized the global discourse through, for instance, the use of the term *bikas*. The global discourse emphasizes specific configurations of ideas and practices of water, at the cost of others.

The identification of domestic water as a basic need in international fora in the early 1970s spurred the establishment of the

DWSS in Nepal. The declaration of the 1980s as the water supply and sanitation decade helped the department expand its projects through funding from multilateral institutions and a host of donors. With the state and NGOs as interlocutors, remote areas in the hills and Terai participated in the global discourse on water, not on their own terms, but according to international agendas. Therefore, the understanding and use of water as a means of subsistence and a medium for ensuring ritual purity is downplayed, as water is seen as a neutral resource used for various purposes, including that of ensuring a healthy life.

The history of water supply in Nepal can also be read as the history of the secularization of water. Earlier, water (*pani* or *jal*), was seen as something that cleaned as well as purified.[16] The organic processes of life such as eating, urinating, defecating, copulating, menstruating, birth and death were perceived as polluting and contact with people involved with them or in these phases of life were said to be polluting, which could be overcome by a ritual bath.[17] The ideology of development – and foreign aid – have increasingly shaped the public perception of water as *jal-srot* ("water as resource"), a neutral medium that promoted cleanliness and health. This led to water being viewed as a resource whose ownership vested with the state, whereas in the past the ownership of water had remained undefined. In short, the ideology of development has led to a significant secularization of water (Sharma 2003).

Studies on the local impact of water projects reveal that while development has secularized water, it has not entirely benefited *dalits* (low castes, outcastes), who continue to be seen as permanently polluting others (ibid.). Dalits were usually left out when formulating and implementing piped water projects, substantiating Pigg's assertion that bikas reinforced existing ritual, social and economic stratifications.

Likewise, the history of water supply in Nepal evinces the diffusion of a technology through social and geographical boundaries. Technological innovations in water supply begun elsewhere and under specific circumstances,[18] first made their way to serve a handful of Rana families and their clientele in Kathmandu in the late 19th century, and over the last century trickled down to much of the peasantry. Still, even at the end of the 20th century, poor peasants and the bulk of the artisan classes, including dalits in Jhirbhanjyang, studied by Sharma (Procuring Water, 2001), still do not have water supply.

The last section looks at why locals aspire to better water supply and sanitation. For those who had some access to water sources earlier, having piped water with a tap-stand or tube-wells, constructed with help from the government (and foreign aid) is a matter of convenience and a status symbol. It is a symbol of modernity, an incentive for locals to participate in the global discourse[19] and be part of bikas. This is as true for a daughter in a household in the Terai hinterlands at the end of the 20th century, as it was for a daughter of a Rana family in Kathmandu, a century earlier.

The perception of tap-stands as a symbol of modernity was not very different from the way European housewives saw indoor plumbing at the turn of the last century.[20,21] Water supply, along with other services, was a consequence of modernity and technological innovations, resulting from the dominance of formal reason. In the late 19th and early 20th centuries, European municipalities competed to expand civic services, including water supply and sewerage, in order to attract business and revenue, which has been well documented by historians.[22] In Nepal, local government bodies jostle for funds from department line agencies and foreign donors, to provide civic services including water supply, electricity, roads and telephones, so their areas can rise in the bikas hierarchy.

In conclusion, it is worth reinstating that a simple thing like domestic water, khane pani, acquires a complexity in the context of development, and means different things to different people. For the state, khane pani is about procuring the hardware and providing water supply; for aid agencies, it is about remaining in the business of acquiring and disbursing funds and improving water supply. Both are crucial interlocutors between the donor headquarters in the first world and the hinterlands of Nepal. For ordinary folk in the hills and Terai of Nepal, it is as much about saving labor and reducing drudgery through this convenience, as becoming modern (*adhunik*).

Notes

1. Dhunge dhara is a traditional stone water spout from where water flows freely while kuwa is a spring water source which may or may not be covered and inar is a man-made well which is usually lined with bricks. Dhunge dhara, kuwa and inar have been the sources through which people have traditionally managed their drinking water needs.
2. Nadi or khola refers to rivers that have water all year round, while kunda is a spring water source with religious significance. The waters of nadi, khola and kunda are used for religious purposes (which could include bathing and oblation).
3. Pokhari is a pond that has still water collected from excess water during the rainy season while khare is a stream that has water only during the rainy season. Kulo is a man-made temporary irrigation canal. All the three, pokhari, khare and kulo are used for the purpose of irrigation (Campbell 1973).
4. By discourse Foucault meant "a group of statements which provide a language for talking about, a way of representing the knowledge about a particular topic at a particular historical moment and refers to both language and social practices Hall, S. and Held, D. (1992). The term discourse as used in this study is informed by Foucault's perspective, which holds that "knowledge" is already deeply invested with power.
5. Writing in the early 1990s and reflecting upon the consequences of development in rural Nepal as of the late 1980s, Stacy Leigh Pigg points to a high correlation between caste status and access to development facilities, where those at the top of the caste hierarchy were also the ones to receive most benefits from development. There are several markers of class differences, and subsequently of stratification, in Nepal: land-owning size, land tenure rights, income, education, etc.
6. The history of water supply and sanitation is limited till 2002, when the Ninth Five-Year plan came to an end.
7. In his presidential address on January 20, 1949, President Harry Truman spelt out the four points that would guide American policies in the years to come. The first related to the interests of the United States, the second to the functioning of liberal market economies, the third to resistance against communism, and the fourth to the problems of underdeveloped areas. The fourth point appreciates the need for sharing technical knowledge, application of modern scientific and technical knowledge and increasing production to ensure peace and prosperity in developing countries. See Escobar (1995).
8. During the first plan, expenditure for drinking water was allocated under public works. From the third to the seventh plans, the table provides sectoral allocations for government expenditures only. In the ninth plan, sectoral allocations are tabulated for development expenditures only.

9. Raimund and Becker-Ritterspach (1995).
10. Stiller and Yadav (1993) pp. 65, 70.
11. Khadka (1991) p. 250.
12. This rationalization began to be increasingly questioned during the early 1990s through studies – some supported by UNICEF itself – which indicated that the link between drinking water and child morbidity was not as simple as first assumed.
13. Although the Asian Development Bank initially concentrated in rural areas, it later branched out into semi-urban areas. From the late 1990s, it was involved in the Melamchi project.
14. The Ministry of Finance (1994), pp. 77–78.
15. See Rural Water Supply and Sanitation Project, Lumbini Zone, Nepal. *Completion Report of the First Phase, January 1990–July 1996, Final Report*, Plancentre.
16. For further discussions on the secularization of water in the context of the West, see Illich, I. (1985).
17. The various religious uses of water in Hinduism is among others, discussed by Kane, P. V. (1976).
18. The reason water supply and sewerage systems developed in European cities in the mid-19th century, was because existing wells and rivers were unable to cope with the ever increasing demographic movements into cities due to industrialization, and which consequently led to the sources being polluted. Besides, there was a growing demand for water from the industries themselves. Cholera epidemics during the early part of the 19th century and the subsequent public health movement in Great Britain led to a growing awareness of the relationship between the quality of water and health and led to the development of water works. In north European cities, development of waterworks was also motivated by the need to improve fire prevention measures (Hietala, M. 1987).
19. This is not to under-estimate the importance of more practical issues like reducing workload, having a year-long more reliable quantum of water, etc. However, abstract symbolism – piped water being a symbol of modernity – does seem to play its role.
20. See Burghart, R. (1993) and Jokinen, E. (1985). A study on consumer's association in dispersed rural areas in Finland in the mid-1980s undertaken by Jokinen, Eeva (1985) entitled *Vesihuollon Merkityksestä* (report 256) published by National Board of Waters, Helsinki (p. 117) further supports this argument. *Vesihuollon Merkityksestä* suggests that the rationale for the expansion of water supply services in rural areas have to do with the understanding of young people living in rural villages that they are modern people, and so deserve to live the same kind of life that are lived in the cities. Consequently the people in the rural areas demand the same services and possibilities that are offered in the suburbs.

21. This analogy between the European housewives at the beginning of the 20th century with rural Nepali women at the end of the century should not be read as suggesting a linear version of history. This is, after all, an analogy, but as an analogy it is apt because both seem to be motivated by similar concerns.
22. Hietala, M. (1987) Op. cit.

References

Burghart, R. 1993. "Drinking Water in the Nepalese Tarai: Public Policy and Local Practice," in Gerard Toffin (ed.), *Nepal Past and Present.* New Delhi: Sterling Publishers.

Campbell, L. B. 1973. *Two Water Case Studies,* Department of Local Development/UNICEF Nepal, Research cum Action Project, Paper Number 5.

Escobar, A. 1995. *Encountering Development – The Making and Unmaking of the Third World,* The Princeton Studies in Culture/Power/History. Princeton. New Jersey: Princeton University Press.

Ferguson, J. 1994. *The Anti-Politics Machine: "Development", Depoliticalisation and Bureaucratic Power in Lesotho.* Minneapolis: University of Minnesota Press.

Hall, S. and Held, D. 1992. *Modernity and its Futures.* London: Polity Press in association with Open University.

Hietala, M. 1987. *Services and Urbanization at the Turn of the Century – The Diffusion of Innovations.* Helsinki: Finnish Historical Society.

Illich, I. 1985. *H2O and the Waters of Forgetfulness: Reflections on the Historicity of "Stuff".* Berkeley: Heyday Books.

Jokinen, Eeva. 1985. *Vesihuollon Merkityksesta,* National Board of Waters, Report 256, Helsinki.

Kane, P. V. 1976. *History of Dharmashastras,* Second edition. Poona: Bhandarkar Oriental Research Institute.

Khadka, N. 1991. *Foreign Aid, Poverty and Stagnation in Nepal.* New Delhi: Vikas Publishing House.

Ministry of Finance. 1994. *Economic Survey 1993/1994.* Kathmandu: His Majesty's Government.

Pigg S. L. 1993. "Unintended Consequences: The Ideological Impact of Development in Nepal," *South Asia Bulletin,* XIII (1 & 2): 45–58.

———. 1992. "Inventing Social Categories Through Place: Social Representations and Development in Nepal," *Comparative Studies in Society and History,* 34 (3): 491–513.

Plancentre. 1997. *Rural Water Supply and Sanitation Project, Lumbini Zone Completion Report of the First Phase, January 1990–July 1996, Final Report.* Butwal: Plancentre.

Raimund, O. A. and Becker-Ritterspach, 1995. *Water Conduits in the Kathmandu Valley.* New Delhi:Munshiram Manoharlal.

Sharma, S. 2001. *Procuring Water: Foreign Aid and Rural Water Supply in Nepal.* Kathmandu: Nepal Water Conservation Foundation.

Stiller, L. F. and Yadav, R. P. 1993. *Planning For People.* Kathmandu: Human Resources Development Research Center.

World Bank. 2007a. *Development and the Next Generation.* Washington D.C.: The World Bank.

———. 2007b. *World Development Indicators,* Green Press Initiative. Washington D.C.: The World Bank.

5

Foreign Assistance, Dependence and Debt

Sanitation Case Study, Kandy, Sri Lanka

Sunil Thrikawala and N.C. Narayanan

The deteriorating financial health of public sector water service providers in developing countries attract large amounts of foreign funds (mostly as loans) and could lead to dependence, deterioration of governance and weakening of institutions. The case study is primarily a whistle-blower to the proposed large-scale Water Supply and Sanitation project to be implemented by the Kandy municipality in Sri Lanka, funded by the Japan Bank for International Cooperation (JICA). The bloated expenditure and high tariff rates to recover the operation and maintenance costs of treating wastewater, could be a huge burden, particularly to lower and middle income households, that raises questions about the larger acceptance of the proposed project. The technological appropriateness, economic viability, social acceptance, political feasibility and thus the sustainability of the project are questioned. The study also explores the related governance questions, in the context of the water sector reforms currently under way in most developing countries, including Sri Lanka.

Introduction

Water supply and sanitation is of growing importance in the current discussions on millennium development goals (MDGs), which in turn attract foreign aid in developing countries, with international financial institutions perceiving huge potential to increase their lending business. In recent decades, the water supply and sanitation policies and programs of most developing countries have been influenced by donor assistance, reflecting strategies of the donor countries and organizations, rather than indigenous strategies of the recipients (Killick 2004; Pradhan 1996). These funding programs can often create dependence with respect to both funding and technology, and might turn out to be inappropriate to the existing institutional structure.

This article focuses on the waste water disposal issues in Kandy, Sri Lanka, that includes a world heritage site because of its cultural, historical and aesthetic values. The city has a population of around 1,60,000 with a water consumption of about 25,000 cubic meters per day. Since there is no proper system of waste water disposal, about 80 percent of this is released into the Kandy lake and Mahaweli river. Over the years, the Mahaweli river, the main source of water in the region, has been threatened with increasing pollutant loads, due to contamination from agricultural and domestic waste, sewage and industrial effluents. Many institutions in different sectors with their own mandates are collectively responsible for this situation.

The National Water Supply and Drainage Board (NWSDB) is the principal authority providing safe drinking water and facilitating the provision of sanitation in Sri Lanka, which began as a sub-department under the Public Works Department, and later became a division under the Ministry of Local Government in 1965. Since 1970, this division functioned as a separate department under the Ministry of Irrigation, Power and Highways, until the NWSDB was established in 1975. At present, the NWSDB functions under the Ministry of Water Supply and Drainage. Consumer metering and billing by the NWSDB commenced in 1982, and several urban water supply schemes operated by the local authorities were taken over by the board, with the intention of providing better coverage and improved service. Rural water supply and sanitation, including deep well programs, are also being implemented by the board (NWSDB Annual Report 2006). The NWSDB has proposed to establish the Kandy City Water Supply Augmentation and Environmental Improvement Project (KCWSAEIP), which is supposed to provide safe sanitation by treating the effluents entering drinking water sources and ensuring clean drinking water within the city. The two purposes of this study are: (a) to examine the financial problems that led the NWSDB to take a loan and (b) to assess the implications of the proposed foreign-aided project, in terms of dependency on technology/expertise and institutional weakening of public systems responsible for the provision of water supply and sanitation services in the country.

The study mostly uses secondary data from local agencies and government institutions. The financial position of the NWSDB was analyzed using some indicators drawn from the data available for the last two decades. The proposed project invited different responses from various sections of society. Key informant interviews were conducted with professionals at NWSDB, Kandy Municipal Council (KMC), the University of Peradeniya, concerned government departments, religious organizations and civil society/research organizations that opposed the project. Since some of the project documents are not in the public realm, and much of the information is confidential, the data was collected through interviews with different officials (thus information triangulated with the university faculty who were involved with an Environmental Impact Assessment (EIA).

The next section discusses the experience and possible implications of foreign aid projects globally and in Sri Lanka, and reviews the debate on foreign aid, aid dependency and development. The third section analyzes the process that led to the deteriorated financial crisis of the NWSDB that forced the organization to go for the proposed project with foreign assistance. The fourth section describes the problem of pollution and suggests solutions while the fifth section analyzes the implications of the project and the larger questions of water governance.

Foreign Aid, Dependency and Changes in Governance

The negative impact of such a project, because of volatility and uncertainty of overseas development assistance flows, fragmentation of donor efforts, project proliferation and complication, conflicting or dominant donor agendas, competition for staff, high administration and oversight cost have already been pointed out (Brautigam and Knack 2004; Knack and Rahman 2004; Birdsall 2004; and van de Walle 2005). Aid practices may impose a substantial burden on qualified public officials, who spend a lot of their time attending to donor concerns and managing aid activities, rather than promoting the development of the country. If aid is not integrated into national budgets, but distributed through parallel structures that cream the best staff from civil service, it can create problems of sustainability. Donors' projects often come with technical assistance that is mostly imposed, without conducting an adequate needs assessment of project design and implementation from the beneficiary's perspective (Mogilevsky and Atamanov 2009). Heller and Gupta (2002) argue that the fiscal uncertainty of dependence on external assistance makes long-term planning extremely difficult.

Aid dependence is defined as a situation in which a government is unable to perform many of its core functions, such as the maintenance of existing infrastructure or the delivery of basic public services, without foreign aid funding and expertise (Brautigam and Knack 2004). Aid also leads to perverse incentives in the making of economic policy. However, there are many other cases, where aid as a subsidy has discouraged collecting revenue, distorted expenditure decisions and undermined the incentives to build state capacity (Moss et al. 2005). Aid projects reduce the revenues to the government by not paying import duties and expatriate personnel are rarely required to pay income tax. The setting up of project implementation units limits the government's ability to learn skills for effective management and administration in aid dependent countries, with donor conditions undermining genuine policy learning (Brautigam and Knack 2004). Capable staff is poached and hived off into parallel structures, weakening institutions by creating resentment and lowering the morale of those left behind (Cohen 1992). In African countries, state capacity has improved little in the last four decades of aid flow and some even point to specific cases of clear decline (van de Walle 2005). We will examine how these arguments could be relevant to Sri Lanka. However before that, let us look at the macro foreign aid scenario in Sri Lanka.

The total commitment of grants and loans made to Sri Lanka till 1999 was US$ 706 million, of which 90 percent were loans. The three major donors, Japan, Asian Development Bank (ADB) and the World Bank, accounted for 77 percent of all aid commitments. Of the loans, 8 percent was for development of the water supply and sanitation sector in 1999 (Foreign Aid Review, Sri Lanka 1999). The government outstanding debt of ₹3,432.4 billion at the end of 2008 is a 14 percent increase over 2007 figures.[1] As a share of gross domestic product (GDP), government debt reduced from 105.4 percent in 2002 to 81.1 percent in 2008 (Table 5.1). Thereafter it increased by 2.2 percent in 2009. The share of government revenue in GDP increased between 2004 to 2006, but has been decreasing since. A similar trend is observed in the share of tax share and government expenditure in GDP.

Table 5.1
Government Debt Indicators in Sri Lanka

Government Finance (percent of GDP)	2002	2003	2004	2005	2006	2007	2008	2009
Government Debt	105.4	105.8	102.3	90.6	87.8	85.0	81.1	83.3
Domestic Debt	45.6	47.9	47.6	39.0	37.5	37.1	32.8	36.5
Foreign Debt	59.8	57.9	54.7	51.6	50.3	47.9	48.3	46.8
Government Revenue	16.5	15.7	14.9	15.5	16.3	15.8	14.9	14.6
Tax Revenue	14.0	13.2	13.5	13.7	14.6	14.2	13.3	12.8
Government Expenditure	20.9	19.0	22.8	23.8	24.3	23.5	22.6	24.9

Source: Central Bank of Sri Lanka, Annual Report 2009.

Financial Crisis in NWSDB

The NWSDB and local authorities like municipal councils are the two major institutions involved in water supply and sanitation services and management throughout Sri Lanka, with the support of the provincial councils, lending institutions, external supporting agencies, community based organisations and non-governmental organizations (NGOs). Water supply and sanitation services were provided by the respective municipal councils until the NWSDB was established as a statutory board enacted by the parliament under the National Water Supply and Drainage Law No. 2 of 1974. Since then, water services were gradually taken over by the NWSDB, and consumer metering and billing commenced in 1982. However, water supply and sanitation services in some cities are still provided by their local authorities, while development activities are undertaken by the NWSDB. For instance, water supply and sanitation services within city limits in Kandy are provided by the Kandy Municipal Council (KMC), while piped water supply to the periphery of the city is provided by the NWSDB. Rural water supply and sanitation, including deep well programs, are also being implemented by the board (NWSDB Annual Report 2006).

The following analysis attempts to assess the financial problems, and consequences, including increased dependence on foreign funding.

Costs and Revenue

Table 5.2 depicts the annual cost of production of water from 2001–07. The cost of production mainly consists of direct operational cost, administrative overheads and other operating expenditure contributing 56 percent, 16 percent and 19 percent respectively, with 8 percent finance cost and 1 percent taxation (NWSDB 2006). The operational expenditure is increasing over the years, with increasing cost of each item for the operations. However, the percentage of cost for each item is either decreasing or not changing, except the personnel cost. Personnel cost as a percentage from total operational cost has increased, despite a drop in 2006. Being an autonomous institution, the NWSDB has the power to increase the water tariff from time to time. In 2009, the tariff rate increased by more than 100 percent. Thus one could argue that the NWSDB was transferring the burden of increased personnel expenditure to the consumers through the increased tariff. This situation, in fact, highlights the necessity of an independent regulatory authority that could control the tariff reforms and other issues of the water board.[2]

Figure 5.1 shows that the development expenditure, revenue generation, customer billing and interest payment are increasing

Table 5.2
Operational Expenditure of Drinking Water over the Years in Sri Lankan rupees

	2001	2002	2004	2005	2006	2007
Personnel	878.24	1012.23	1919	2541	1829.35	2335.09
	(39.09)*	(35.39)	(45.61)	(48.19)	(44.83)	(47.63)
Pumping	904.8	1135.42	1126	1245	1360.67	1587.87
	(40.27)	(39.69)	(26.76)	(23.61)	(33.35)	(32.39)
Chemicals	170.33	182.24	171	214	319.57	349.49
	(7.58)	(6.37)	(4.06)	(4.06)	(7.83)	(7.13)
Repair & Maintenance	130.07	152.94	224	308	256.86	287.03
	(5.79)	(5.35)	(5.32)	(5.84)	(6.29)	(5.85)
Establishment	68.28	272.57	451	610	137.64	152.01
	(3.04)	(9.53)	(10.72)	(11.57)	(3.37)	(3.10)
Other	95.15	104.98	316	355	176.44	191.04
	(4.23)	(3.67)	(7.51)	(6.73)	(4.32)	(3.90)
Total	2246.87	2860.38	4207	5273	4080.53	4902.53

Source: NWSDB Annual Reports, various years (2001–07).
Note: *Values in parenthesis are the percentage costs of each item from the total cost.

Figure 5.1
Total Revenue, Development Expenditure and Loan Interest Over the Years (in Sri Lankan rupees—SLR)

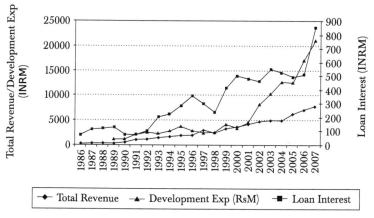

Source: NWSDB Annual Reports, various years (1987–2006).

over time. However, development expenditure has overtaken the revenue and customer billing, except on a few occasions. Interest payments were increasing rapidly since 1998.

Non-revenue Water

The production of piped water has been increasing over the years. However, non-revenue water (mostly water leaked between the delivery point and consumer meters[3]) continues to be around 30–40 percent, though there is a slight decline as is visible in Table 5.3. Thus, NWSDB was unable to account for 140 million cubic meters of treated water, costing more than ₹2,300 million (INR or SLR) during 2007 (Table 5.4). Non-revenue water in 2007 was sufficient to supply water to approximately 585,000 new households, assuming an average consumption of 20m^3 per household per month. The operations and maintenance cost can also be substantially reduced by cutting down the non-revenue water.[4] This shows the discrepancy between production and distribution, and widening gap between water billing receipts and

Table 5.3
Piped Water Production and Non-revenue Water Over Time (all SLR)

Year	Piped Water Production (mm³)	Non-revenue Water (%) Greater Colombo	Other Areas
1990	219	36	40
2000	332	39	31
2001	343	39	30
2002	350	37	30
2003	357	38	31
2004	367	36	29
2005	383	35.9	33.8
2006	398	37.5	34.4
2007	424	37.8	33.1

Source: NWSDB Annual Reports, various issues (1987 to 2006).

Table 5.4
Non-revenue Water and Costs Incurred

	2003	2004	2005	2006	2007
Non-revenue water (000m³)	124,442	123,959	96,690	136,630	140,491
Direct cost (million rupees)	1406.5	1648.4	1290.1	2040.4	2377.9
Additional cost borne by consumers for every unit of consumption (₹)	6.04	6.76	6.76	7.82	8.37

Source: NWSDB Annual Report 2007.

the total revenue, indicating lowering of institutional capacity over the years.

The non-revenue water (Sri Lanka or Kandy) at present is around 38 percent, and the KMC is satisfied to have it brought down to 42 percent. They have identified the sources of non-revenue water leakage (Table 5.5) and further research is needed to identify the most significant source before taking any action to cut down its leakage. However, there are few difficulties in cutting down non-revenue water enlisted as follows:

Table 5.5
Types of Non-revenue Water

Type of Loss		
Unbilled authorized consumption	With meters	Government schools and KMC buildings
	Without meters	Public stand posts, water supplied by tanks
Apparent losses	Unauthorized consumption	By-passed meters
		Defective meters
	Consumer meter inaccuracies	Meters with errors
		Deliberate damage to the meters
		Improper data storage
Real losses	Damages in main pipes	Internal leakages at main pipes and valves
	Overflows and leakages in storage tanks	
	Leakages in connection pipes	Leakages in pipes and valves in private connections

Source: Adopted from Kamani et al. 2012.

(a) Authorities are not informed of the leakages immediately.
(b) There are not enough water storage facilities at individual units, though it is a requirement for a water connection, particularly in the high density towns. So the KMC has to wait until night to temporarily shut down the water supply for repairs.
(c) Difficulties in accessing the damaged infrastructure due to ad-hoc construction within the city.

Water Treatment Costs

Pollution by sewage, fertilizers, pesticides and toxic metals degrade the quality of water available for human consumption combined with NWSDB incurring progressively higher cost in treating water (see Table 5.6). In Sri Lanka, water is treated at the full treatment level, in which the processes of screening, coagulation, flocculation, sedimentation, filtration and disinfection are involved. The processes which cost INR 60,000 per connection alongwith an operation and maintenance cost of INR 10–25 per cubic metre (Attanayake and Athukorala 2007). There is a huge social cost incurred due to the irresponsible use and pollution of water bodies.

Foreign Aid for NWSDB

In 2007, the NWSDB obtained SLR 14,166 million as foreign funds and the government of Sri Lanka provided 4,587 million SLR as counterpart funds (NWSDB Annual Report 2007). The NWSDB is handling 15 foreign-funded projects, with funding mostly from the ADB, Japan International Cooperation Agency (JICA) and German Agency for Technical Cooperation (GTZ). Figure 5.2 shows a sharp increase in foreign funds coming as grants and loans during 2000–07. In 2005, the NWSDB obtained SLR 36,338 million as a foreign grant and SLR 12,841 million as a loan (Annual Report 2006). Even though the revenue income was increasing along with the increasing service connections by 2004, the Board had a deficit of over INR 800 million ,that came down to SLR 22 million, due to improved revenue and reduced interest paid in 2005. The level of high interest on loans, along with high level of foreign grants and loans, indicate the dependency of the

Table 5.6
Water treatment costs

Treatment	Unit Involvement	Cost of Treatment Per Connection (INR)	Operation and Maintenance Cost Per cu.m (INR)
Only disinfection	Screening and disinfection	1,500	0.5–1.5
Minimum treatment	Screening, filtration and disinfection	8,000	1–3
Partial treatment	Screening, 'roughing Filtration', filtration & disinfection	16,000	3–10
Full treatment	Screening, coagulation, flocculation, sedimentation, filtration and disinfection	60,000	10–25
Advanced treatment	Screening, coagulation, flocculation, sedimentation, filtration and adaptation, and disinfection	150,000	75–100

Source: Attanayake and Athukorala 2007.

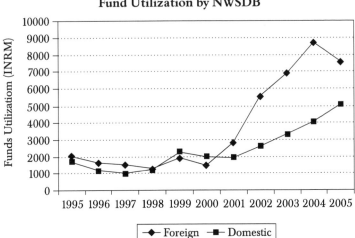

Figure 5.2
Fund Utilization by NWSDB

Source: NWSDB Annual Reports, various issues (1987–2007).

water board on foreign funds. Further, development expenditure exceeded total revenue, except on a few occasions, indicating its dependency on external funds. In 2007, the foreign fund allocation was more than double the local fund allocation. The provision of water services is increasingly dependent upon foreign support, as is evident through ever increasing development expenditure and the widening gap of foreign funds to local funds.

Waste Water in Kandy: Problem and Proposed Solution

The Mahaweli is the most important river for Sri Lanka, as it provides water for multiple purposes such as domestic and industrial use, hydropower and irrigation. Over the years, the Mahaweli river has been threatened with increasing pollutant loads due to contamination from agricultural, industrial and domestic waste and sewerage. Urbanization and increased population in Kandy city, coupled with inefficient sewage systems, contribute heavily to the pollution of the Mahaweli river. The sewage in the downtown area flows directly into the drainage system. Further, sewage—mainly gray water from hotels, offices and residences around the Kandy lake, flow into the Kandy lake, causing algae bloom. Kandy has a long history of indigenous design and management of drinking water. The current problem in Kandy city and the expected benefits and estimated costs of the Kandy City Water Supply Augmentation and Environmental Improvement Project (KCWSAEIP) are discussed in this section. This includes the technology/expertise involved, impact on the institution and the sustainability of the proposed project.

The NWSDB is planning to operate a Sewage Treatment Plant (STP) in two phases with a capacity of 8500 cubic meter per day. The project plans to collect waste water from Kandy city through pipelines, pump it to a treatment plant, treat it and dispose the treated effluent into the Mahaweli river.

The expected benefits are as follows:

(a) Reduction in public nuisance due to increased safe water sources and reduction of open-air sewers in the urban area.
(b) Improvement in public health resulting from the reduction of water-borne diseases.
(c) Improvement in surface and ground water quality.
(d) Elimination or minimization of pollution in Kandy Lake.
(e) Improvement of the aesthetics of Kandy Lake area.
(f) Reduction in odor problems in the city.
(g) Improvement in the raw water quality at the Gohagoda intake[5] in the Mahaweli river.

The project is estimated at almost INR 14 billion, of which 82.9 percent is coming as a soft loan through the treasury by Japan Bank for International Cooperation (JBIC); (see Table 5.7) and the estimated operation and maintenance cost is going be INR 11 million per month, to dispose of 17,000m^3 of waste water per day for 55,000 people in Kandy City.[6] This works out to INR 210 per month per household for waste water treatment alone.

From the foreign portion of the loan, which includes construction contracts, provisional sums, contingencies and engineering services, almost 50 percent goes back to the lenders. The sustainability and even the initiation of the system after completion depend on the Sri Lankan government, exclusively spending huge

Table 5.7
Estimated Costs of the Sewage Treatment Plant

Item	Foreign Cost Million Yen	Local Cost Million SLR	Total Cost (Million SLR[7])
Construction contracts	4,352.2	3,450.9	7,407.4
Provisional sums			
Road reinstatement	0	334.0	334.0
Power supply	0	10.0	10.0
Equipment procurement	133.1	0	121.0
Sub total	133.1	344.0	465.0
Contingencies	224.3	2,515.1	2,729.0
Engineering services			
Detailed design	318.7	145.6	435.4
Construction management/supervision	255.9	152.5	385.1
Price contingency on engineering services	129.6	0	117.8
Sub total	704.2	298.1	938.3
JBIC loan portion	5,413.8	6,609.1	11,539.7 (82.90%)
Project administration	0	111.1	111.1
Land acquisition	0	0	0.0
Custom duties	0	539.4	539.4
Value added tax	0	1729.4	1,729.9
Government of Sri Lanka portion	0	2379.9	2,379.9 (17.10%)
Total Cost	5,413.8	8989.0	13,910.6

Source: SAPROF Report 2005. Provided by the authors.

sums on household connections and ensuring power supply. The responsibility of inclusion of low-income settlements and public sanitation also rests exclusively with the government, with no donor spending on this. Road repairs after the pipes are installed are also the responsibility of the government. The cost of engineering services and expertise from Japan, and the huge benefits of selling the (black box of) technology for the plant and pumping station as seen in the Table 5.7, clarifies the unequal benefits and costs accrued by the "donor" and "receiver." The provisional sums of equipment procurement and engineering services consist of detailed designing, construction, management, supervision and price contingency on engineering services. Thus of the SLR11.5 billion rupees that comes as the loan, a major share will go to the donors, with questions of sustainability at various levels (as will be discussed in the sections to follow). With delays in the completion of such projects, the escalation of costs could be a huge burden to the exchequer. An important observation here is the escalation of the total project cost from SLR 3 billion in 1998[8] to the current estimated project cost of INR 18.11 billion (Yen 14.09 billion), as per the agreement signed between JICA and Government of Sri Lanka on March 21, 2010.[9] With the ongoing delays, we can easily anticipate further escalation of costs.

Implications of the Project

The implications of the project are evident in Figure 5.3. The project is expected to source funds from a foreign donor and needs foreign technology to be imported, thus, creating dependency. Bringing in new technology along with policy changes has to be suitable to the existing institutional set-up, especially qualified staff who can handle the new technology. Either the existing institution has to increase its technical capacity, or affiliate with other institutions. This will have an impact on the project cost

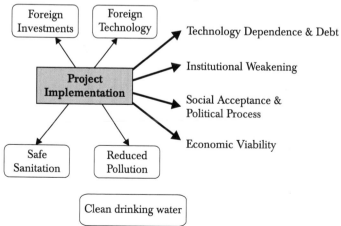

Figure 5.3
Kandy City Water Supply Augmentation and Environmental Improvement Project and its Implications

Source: Provided by the authors.

and the hence sustainability. New skills and the funds will have an impact on the organizational culture. Developing a new work culture may be detrimental to the existing set-up. When the new project is implemented, it has to recover at least the operation and maintenance expenditure, which it usually does through increased water tariffs. If the tariff is unaffordable, it will lead to exclusion. It is also important to look at whether there is a political will to implement the project and the alternative scenario if the project is not implemented. The discussion so far justifies the exploration for alternative lower cost technologies by mobilizing indigenous resources.

Technological Appropriateness

Feasibility of a centralized system

The technology for waste water treatment has come as a package with the loan from the donors. The nature of technology, its suitability for local circumstances, cost of alternative technologies etc. still remains to be tested. The proposed treatment technology – using an "oxidation ditch" – is highly sophisticated and complicated. An environmental impact assessment done by a private firm has cleared the technology, claiming that under prevailing conditions (continuous increase in population, limited land availability, soil type and high water table in the area), a centralized waste water treatment method is needed to avoid contamination of the Mahaweli water by sewage (NRMS 2005). However, the report does not analyze any alternative technologies. Further, a restructuring of Kandy City has already been approved with several big polluters like the prison, bus station, etc. to be relocated outside the city.[10] In that case, the space problem highlighted as the justification for the centralized system become insignificant.

Organic Pollution Load and Uncertainty

According to the comments submitted by the Environmental Foundation Limited to the Central Environmental Authority, the effluents from the proposed plant will increase the organic pollution load by up to 700 percent, even though it is actually supposed to reduce the pollution level. However, according to

Central Environmental Authority standards, if the Biochemical Oxygen Demand (BOD) level can be maintained below 20mg/l, increase of organic pollutant loading does not harm the water quality because of the self-oxidation ability of moving water. The Environmental Foundation Limited also argues that during periods of plant failure (which are quite unavoidable according to the past experience of small scale sewage treatment plants in the local context), it would increase the organic pollution load up to 4.2 tons of BOD per day (NRMS 2005). The heterogeneous composition and unpredictable in-flows of waste water is difficult to handle, and hence the only reasonably successful plants in Sri Lanka are the common sewage treatment plants of industries, in which the volume and composition of the load from each industry is predictable and constant.[11]

Project Dependency, Lifespan, Operations and Maintenance

After such large investments, if this is unavoidable, donors may send you huge bills for maintenance in future (elaborated later). The expected lifespan of the project is 30–40 years (NRMS 2005). However, this is possible only if appropriate operation and maintenance activities at the sewage collection and treatment plant are maintained. The sustainability of the system rests on a set of sequential activities that demand experts, technicians and skilled labor. Shortfall in any activity could cause the entire system to collapse, as seen in similar, but smaller scale plants.[12] Poor maintenance was a major problem. The operation of the sewage treatment plant entirely depends on electric power for running pumps at various locations. Power failure at any location could cause the entire system to shut down. The risk of blocking the sewerage in-flow, or power failure at the sewage treatment plant, could cause spills and overflow of tanks, leading to sanitation and health hazards and odor problems. Since power failures are very common, the subsequent chaos can be frequent. This implies that the proposed technology might increase the risks and health hazards that it is supposed to curb. Further, the pipe system has to be laid in a very systematic manner to collect and pump the sewerage. However, Kandy is not a planned city as such, and all the roads, buildings and other structures have been built in an ad-hoc manner. Thus, setting up a systematic pipe system for sewerage can be quite challenging.

Economic Viability

Enhanced Debt

The project is estimated at almost INR 14 billion, of which 82 percent is funded by the Japan Bank for International Cooperation, with the estimated operation and maintenance cost being INR 11 million per month to dispose of 17,000m^3 per day to serve 55,000 people in Kandy district. The money is coming as a soft loan from the Japan Bank for International Cooperation, to the treasury, and as a grant from the treasury to NWSDB. It is 59 percent of the GDP (GDP = SLR 2,350 billion[13]) and the current debt of each individual will go up by 0.12 percent to serve a very small fraction (0.3 percent) of the total population of Sri Lanka. This is in fact dependency on money from foreign donors.

Huge Operation and Maintenance Costs and Increased Tariffs

The NWSDB is planning to hand over the plant to the KMC once the construction is completed. The KMC will be the local authority responsible for cost recovery and maintenance. The environmental impact assessment report (NRMS 2005), as well as Table 5.7 indicates that most of the construction and materials will have to be imported along with foreign expertise. The sustenance of the system needs continuous import of materials for repairs and replacements. Thus NWSDB and the KMC will be continuously dependent on the donors' technology.

The KMC is planning to recover the full operation and maintenance cost for proper maintenance of the project. This is in addition to the asset management costs, that work out to 10 percent of the capital costs for repairs and replacement.[14] The water board has understood that adding this to the water bill is going to be a huge financial burden for the users and will be a major reason for possibly long term dependency on external sources for technology and funding.

The operation and maintenance consists of personnel, electricity, chemicals, repairs and maintenance, and establishment costs, as given in Table 5.8. The total operation and maintenance cost is INR 101.3 million per annum. The per capita sewerage flow has been estimated at 86 l/day, which is equivalent to 0.086m^3/day (NRMS 2005). If we assume a five-member family, there will be 0.43 m^3/day of sewerage discharge per household per day. The total operation and maintenance cost of disposing off 17,000 m^3/day works out to INR 16.33/m^3. Then each household will have to pay INR 210/month for the service. However, the water board is proposing a differential tariff system based on household income and the type of users. On average, a medium income group household will have to pay INR 200/month.[13] In addition, a connection charge, which depends on the distance from the household to the main sewage pipe, is another financial burden for the consumers, with very few people willing to get the connection, even if a fully reliable service is guaranteed. More than 40 percent of the total cost is taken up by chemical repair and maintenance, by which at least 90 percent of the funds go back to the donors annually, and the water board and KMC will forever be dependent on foreign replacement of parts and chemicals. The environmental impact assessment report indicates that most of the equipment will have to be imported (NRMS 2005).

Table 5.8
Direct Operation and Maintenance Costs of Kandy City Water Supply Augmentation and Environmental Improvement Project

Item	Operation and Maintenance Costs (Million Rupees/annum)
Personnel	8.8
Electricity	38.8
Chemicals	16.7
Repairs and Maintenance	34.6
Establishment	2.4
Total	101.3

Source: Feasibility Report of the Kandy City Water Supply Augmentation and Environmental Improvement Project.

Institutional Weakening and Dependency

The construction of the plant, operation and maintenance, and repairs need experts and skilled workers, for which there is a plan for allocating 27 personnel for operation and maintenance for various tasks such as inspection, sewer pumping, sewer treatment, water analysis, etc. NWSDB and the KMC will have to depend on external expertise during the construction, operation and maintenance and in emergency breakdowns as well (NRMS 2005). The project has no budget allocation for capacity development of NWSDB personnel, hence the technology and processes become a problem for the institution, often leading to hiring of over-priced foreign and local consultancy expertise that results in overspending. The lack of institutionalization of expertise by NWSDB will lead to de-skilling and enhancement of costs. The opaqueness of the project office regarding sharing of vital information does not allow us to assess this aspect in all its detail.

Social Acceptance and Political Processes

The environmental impact assessment study (NRMS 2005) has conducted two surveys to determine the opinion of the people living around the proposed sewage treatment plant site, and the general public of Kandy City. The survey around the plant site was conducted with 165 respondents of which 68 percent supported the project, since they believed that the project would ensure cleanliness of Kandy City, reduce environmental pollution, meet the needs of the increasing population, maintain the water quality in the Mahaweli river and scope of recycling waste for agricultural production; 40 percent opposed[15] the project, citing uncertainty in maintenance, resulting in bad odor, adverse impact on the botanical garden due to gas emissions, proximity to the Buddhist temple, treating/reusing water with human excreta and the negative impact on the experimental fields of the Department of Agriculture.[16] The survey concludes that the majority of the people support the project with genuine concerns about possible bad odor, poor maintenance and apprehension of additional payments (NRMS 2005: 103–6).

There were protests about the project, with over 5,000 people participating. The protests centered around the site for the sewage treatment plant, organized by the Gatambe Buddhist Temple with the participation of monks, religious leaders, officials of Department of Agriculture and the general public. The protest questioned the survey findings with its "manufactured consent". It was argued that there was no social acceptance for the project, and the lack of transparency hid the implications of the project. If a genuine public debate was held, with a transparent sharing of information about the technological, economic and institutional implications, it is doubtful whether it would pass the test of democratic consensus to take the project forward.

Conclusion

In the context of the financial crisis of Sri Lanka's NWSDB, this study attempted to investigate donor influence and institutional changes in the proposed foreign-aid project in Kandy. The techno-economic analysis brought out the implications for dependency and debt. Insights from literature about the need for strong policies and institutions to have aid-effectiveness were validated in this case. The unequal terms of the aid, and weak bargaining power of the recipient, would breed technological, knowledge and economic dependency on donors. The price tag for the engineering services and expertise that come from Japan and the huge benefits of selling the (black box of) technology for the plant clarifies the unequal benefits and costs accrued by the "donor" and "receiver." The proposed technology might not be the most appropriate, with very high capital cost, operation and maintenance cost and asset management cost, where the last two are recurring and essential for the sustainability of the plant.

The decentralized technological alternatives like (a combination of) individual septic tanks and constructed wetlands could be integrated at the mini/micro watershed level, to be managed by community-based institutions. Future studies have to streamline the technological alternatives to suit diverse biophysical and socio-economic contexts with appropriate institutional arrangements. Further, a thorough economic analysis should be done to investigate the possibility of establishing decentralized individual systems of sewerage management. Between the two approaches of supply-driven urban water supply and sanitation, it is worthwhile to examine the possibility of practising a demand-driven community-based approach through mobilization of community resources. However, experience in recent years caution about a fully "autonomous" way of community-driven development for water supply and sanitation that imposes a financial burden on the poor because of the high operation and maintenance costs and low capacities at the local level, that threaten the sustainability of such decentralized systems. This is where civil society could facilitate state-community (private-public) partnerships for effective synergy in governance of water supply and sanitation systems.

The concluding message is not a plea against all donor funding, since Sri Lanka, like many other developing countries, is dependent on aid. Paradoxically, the alternatives are also foreign-funded. However, they have built-in resilience mechanisms of simple, locally appropriate technology, banking on local labor and pooling together local resources, both financial and human. It has institutional mechanisms for a demand-driven approach that builds on consensus and ownership with the participation of local NGOs and Community Based Organizations (CBOs). All these might throw up other questions, which we have raised elsewhere (Narayanan and Mohammed 2007). But, in comparison, the alternatives offer scope for less dependent (technologically and financially) solutions for water supply and sanitation problems in Sri Lanka and elsewhere.

We will conclude this article by raising two larger questions that came up from earlier discussions—the first is why is there a policy divergence between the water supply and sanitation service provision between rural and urban areas, with a completely demand-driven and community-operated approach in rural areas and a supply-driven and state-operated approach for urban areas? In a welfare-oriented paradigm, the reverse is needed, where the poor rural population has to be supported by state provision of services. An easy and management-oriented justification would be that urban services open up possibilities of building up water supply and sanitation infrastructure that maximize benefits for the largest population in a confined space compared to the sparse and geographically distributed rural population. A less obvious answer is that the political economy of policy formulation favors the urban rich that corners state subsidies for service provision. A further question then is whether in vital sectors like water supply and sanitation, the state ought to be rather the provider than facilitator of services, since it has huge implications for the health of the populace. These are relevant governance questions that ought to be discussed in policy debates in future.

Notes

1. Available at http://www.cbsl.gov.lk/pics_n_docs/_cei/_docs/ei/wei_20090130.pdf (accessed September 20, 2010).
2. Currently, the only regulatory mechanism prevails on the NWSDB is that it has to present its annual report to the Commission of the Public Enterprises (COPE).
3. Other types of non-revenue water are unauthorized water usage, errors in measurement at different locations, inaccurate meter readings, defective meters, unbilled legitimate consumption, etc.
4. Tortajada (2006) estimated that for the year 2001 the O&M cost/m3 of drinking water could have been reduced to INR 13.00 from INR 20.20 if the NRW was eliminated.
5. This is the current intake of water from the Mahaweli river for city water supply. This intake is 500m downstream of the proposed STP outlet to the river.
6. Available at http://www.jica.go.jp/srilanka/english/office/topics/press 17_01.html (accessed September 22, 2010).
7. Exchange rate: 1 SLR = 0.91 Yen.
8. http://www.kandynews.net/august_2007/news.shtml.
9. Available at http://www.jica.go.jp/english/operations/evaluation/oda_loan/economic_cooperation/pdf/sri100326_02.pdf (accessed September 20, 2010).
10. http://www.sundayobserver.lk/2010/01/24/new41.asp. (accessed on January, 2010).
11. Personal Communication with Mr M. Sivakumar, Former Deputy Director of Northern and Eastern Province of Central Environment Authority on September 27, 2007.
12. The sewerage treatment plants at Peradeniya Hospital, Hantana housing scheme, Raddoluwa housing scheme all failed.

13. Source: Economic and Social Statistics of Sri Lanka 2006, Central Bank of Sri Lanka.
14. Personal communication with Mr J. C. Jayalath, Project Director, STP Project.
15. The figure does not add up to 100 percent because some people had mixed opinions (NRMS 2005: 103–4).
16. A larger survey was done covering the benefiting areas in Kandy city with a sample of 100 households. There were 78 percent, 3 per cent and 19 percent supporting, opposing and with mixed opinions. Apart from the reasons given above, improved health conditions of the public, creation of job opportunities and efficient usage of small plots of land were reasons given by those supporting the project.

References

Attanayake, M. A. M. S. L. and K. Athukorala. 2007. "An Experience of an Epidemic: Highlighting the Need for Collective Decision-making through IWRM Based Strategy." Paper presented at the Second South Asian Research Workshop on Water Supply Sanitation and Wastewater Management. 24–26 September, Kandy, Sri Lanka.

Brautigam, N. and S. Knack. 2004. "Foreign Aid, Institutions and Governance in Sub-Saharan Africa," *Economic Development and Cultural Change*, 52(2): 255–85.

Birdsall, N. 2004."Seven Deadly Sins: Reflection on Donor Failings." CGD Working Paper No. 50, The Center for Global Development, Washington D.C.

Central Bank of Sri Lanka. 2006. *Annual Report*. Sri Lanka: Central Bank of Sri Lanka.

Cohen, J. M. 1992. "Foreign Advisors and Capacity Building: The Case of Kenya," *Public Administration and Development*, 12(5): 493–510.

Heller, P. and S. Gupta. 2002. Challenges in expanding Developmental Assistance. IMF policy discussion paper.

Killick, Tony. 2004. "What Drives Change in Ghana? A Political Economy View of Economic Prospects." Mimeo.

Kamani, H., M. Malakootian, M. Hoseini and J. Jaafari. 2012. "Management of Non-Revenue Water in Distribution Network and Conveyor Lines: A Case Study," *Journal of Health Scope*, 1(3): 147–152.

Knack, S. and A. Rahman. 2004. "Aid Dependence and the Quality of Governance: A Cross-country Empirical Analysis." Background Paper, World Development Report.

Moss T., G. Pettersson and N. van de Walle. 2005. "The Aid Institution Paradox: Long-term Cost of High aid Dependency. Available online at http://www.cgdev.org/content/publications/detail/5646 (accessed on July 30, 2007).

Mogilevsky, R. and A. Atamanov. 2009. Technical Assistance to CIS Countries. CASE Network E-briefs, http://www.case-research.eu (accessed on November 10, 2013).

Narayanan, N. C. and Irshad Mohammed. 2007. "Foreign Assistance and Institutional Weakening: Experience of Provision of Drinking Water in Kerala, India." Paper presented at the Workshop on Civil Society in Water Governance, 10–12 October, Mulshi, Pune, India.

National Water Supply and Drainage Board (NWSDB). 2006. *Annual Report*. Sri Lanka: National Water Supply and Drainage Board.

Natural Resources Management Services Private Limited (NRMS). 2005. "EIA Report of Kandy City Waste Water Disposal Project: The Environmental Impact Assessment," Central Environmental Authority, Sri Lanka.

Pradhan, S., 1996. "Evaluating Public Spending: A Framework for Public Expenditure Reviews," World Bank Discussion Paper No. 323, World Bank: Washington D.C.

van de Walle, N. 2005. *Overcoming Stagnation in Aid-Dependent Countries*. Center for Global Development, Washington D.C.: Brookings Institution Press.

# 6

Institutional Changes, Public Provision and Drinking Water Supply in Kerala

N. C. Narayanan and *S. Mohammed Irshad*

Overview

In the last two decades, the water sector, especially drinking water policies and programs in India witnessed profound changes following the sector reforms described in Chapter 1. Correspondingly the move has been from a supply side towards demand side water management (Sangameswaran 2010). The basic premise for doing so is the perceived inability of the government to effectively administer the existing water infrastructure, provide water in an economically efficient manner to users and muster further financial resources to infuse in the water infrastructure (Cullet 2006). Regarding the rationale of such reforms, there are differing points that view these either as a response to the failure of the welfare state and of top-down technocratic approaches that require change or a more critical approach that views the reforms as a direct outcome of the pressures of international financial institutions, and of the influence of neoliberalism, or as involving the role of changing rural and urban elites (Sangameswaran 2010). Multilateral institutions (such as the World Bank, DFID) suggested that the state lacked capacity for long-term vision and thus, future lending be made conditional on major long-term policy changes in the sector. The prescriptions were manifold like divesting the state from responsibilities other than policy making; engaging the private sector for provision of services and establishing independent regulatory agencies (Wagle and Dixit 2006). The implementation includes legal, institutional, financial and regulatory changes that focus on financial sustainability and technical solutions to the existing problems in the water sector and include suggestions for unbundling of tasks, increasing tariffs, cost recovery for financial sustainability, elimination of subsidies, retrenchment, privatization and public-private partnerships (PPPs) and allocation of water base on market principle (Dharmadhikary and Dwivedi 2009). The main thrust of water sector reforms is to transform the role of the government by transferring a part of the existing governmental prerogatives to users and private actors and also to set up new bodies at the local and state level to take over part of the functions of the government (Cullet 2006). This shift from "government" to "governance" has been associated with the dilution of the role and importance of the state—the state progressively lost some of its central functions to supranational, infra-national levels, and non-state institutions, such as private companies and voluntary organizations (Jessop 1994). Governance thus becomes a process of sharing social, political, administrative aspects, where the state, market and civil society have their own roles, without the state having a central role, since no single sphere is supposed to have sufficient knowledge to dominate the governing model (Rhodes 1997). In this liberal version of governance, institutions are created for coordination among different members of the society, and to prevent conflicts of interest, by defining the rules of the game (Gorringe 1997). It also represents a rediscovery of the role which civil society and the institutions can play in promoting collective private and public ends (Mackintosh 1992). Although not our central focus, we shall explore the implications of such a liberal version of governance in this article. Our major presumption is that what is missing in this literature and larger debate is an exploration of the specific constraints of public utilities responsible for provision of drinking water — aspects of the so called "state failures," an assessment of the emerging institutional alternatives and its implications.

This study examines Kerala's recent experience of drinking water governance, especially the institutional shifts with some foreign assisted projects for the provision of drinking water. The inquiry presumes that such interventions induce governance changes, especially the waning role of state in the provision of vital services like drinking water. Kerala has a history of a reasonably vibrant public space and institutions, and hence development NGOs have traditionally had a relatively minor role, unlike their counterparts elsewhere in the country. Recent changes have seen the emergence of para-state bodies and government-organized NGOs in the provision of water services. The study raises the question whether such a process will lead to a weakening of the public institutions responsible for drinking water provision that might have a differing impact on various societal groups regarding access and control over water.

A backgrounder with a brief historical analysis of the evolution of water supply and the recent institutional changes has been made from available literature and key interviews. The third section discusses the history of Kerala's public sector water provision and the financial crisis that compelled the Kerala Water Authority (KWA) to go in for foreign assistance. This is analyzed by budget and annual accounts documents of KWA of the last one to two decades. The subsequent sections deal with the resulting institutional changes and its implications.

Kerala Water Authority: State to Para-state and Continuing Financial Crisis

State to Para-state

The erstwhile princely state of Travancore, initiated the Willingdon Water Works in 1931 to provide water supply to Thiruvananthpuram, the capital city of Kerala, for a prospective population of 175,000. The project was named after Lord Willingdon, the then Viceroy and Governor General of India. The project was implemented by the Water Works and Drainage Engineering Department at a capital cost of INR 5.6 million, with the Thiruvananthpuram Municipality contributing annually towards the working expenses, which it met by levying a water tax.[1] After independence in 1947, the water supply service was made part of the Public Health Engineering Department of the Public Works Department (PWD), Government of India. This followed the realization that 55–60 percent of the diseases are water borne or transmitted due to lack of proper sanitation facilities. In 1956, a the Public Health Engineering Department (PHED) was formed under the Health Department as per another instruction of the Government of India. After the formation of Kerala state in 1956, public health divisions were formed in all districts. In 1971, the administrative control of the PHED was transferred from the Health Department to the Local Administration Department. The PHED initiated large scale piped water schemes in Kerala and was brought under the newly-formed Water Resources Department in 1982, and converted into an autonomous organization called the Kerala Water and Waste Water Authority in 1984 by an ordinance of the Government of Kerala. The Kerala Water Authority (KWA) was formed through the Kerala Water Supply and Sewerage Act of the state legislature, passed on July 29, 1986. The major objective of forming such an autonomous body was to ease the financing and management of water supply and sanitation-related activities. These shifts could be related to the changing policy of the Union Government favoring transformation of public sector enterprises to autonomous entities,[2] and a World Bank loan that was the first foreign aid to the KWA.

Financial and Administrative Changes in KWA

The KWA is the only public sector organization engaged in harnessing and distributing piped drinking water in Kerala, with a huge installed capacity. It also claims a history of sustainable utilization of the available surface water sources[3] and technical expertise to extend new services at reliable cost.[4] Yet, this built-in capacity of the authority seems to be under stress, since like other public sector utilities, the KWA is also reeling under a financial crisis.

Figure 6.1 explains the sources of funds and trends for drinking water in Kerala. It is clear that state plan expenditure is on the decline, with an increasing dependence on external (foreign) funds

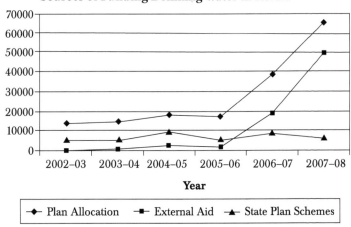

**Figure 6.1
Sources of Funding Drinking Water in Kerala**

Source: Economic Review, Government of Kerala, 2008: Appendix 5.15.
Note: State Plan Schemes include the Government of Kerala's own financing. The total plan allocation includes state plan schemes, Life Insurance Corporation/Housing and Urban Development Corporation, externally aided project funding, central government assistance, etc.

and other sources, including central government funds. Figure 6.2 assesses the financial health of KWA through four components– interest payments by KWA, revenue receipts, Government of Kerala grants to KWA and capital expenditure by KWA. Since 1993–94, interest payment incurred by KWA is higher than the other three components, and the capital expenditure is the least, with all the four components increasing at higher rate from 2003–04.

The trend of capital expenditure of rural and urban water supply projects is on the decline as shown in Figure 6.3, which raises a question mark on the sustainability of infrastructure. Higher interest payment reflects the higher debt profile of KWA. Hence, every additional investment necessarily leads to further borrowing, undermining the institutional capacity of the authority.

Further, the urban water supply investment is less than that in the rural sector since 1985–86, except from 2003–04, when urban expenditure began overtaking rural expenditure. Moreover, it is seen that from 1985–86 to 1997–98, the rural capital expenditure

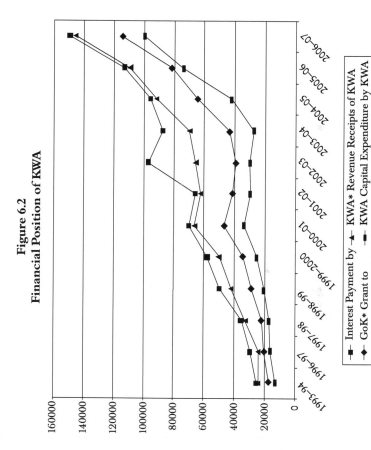

**Figure 6.2
Financial Position of KWA**

Source: Annual Accounts of KWA (1993–94 and 2006–07).
Notes: *Kerala Water Authority; **Government of Kerala.

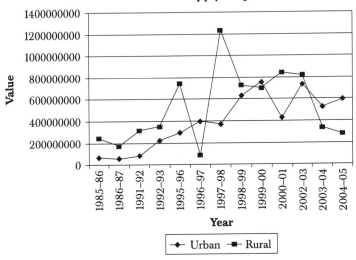

**Figure 6.3
Capital Expenditure on Urban and Rural Water Supply Projects**

Source: Annual Accounts of KWA (from 1985 to 2004).

had been higher than urban capital expenditure. Investments in rural drinking water schemes by KWA since 2000 have declined, and it is attributable to the Jalanidhi project that mobilized massive foreign assistance exclusively for the rural sector. The financial problems forced KWA to approach the World Bank and Japanese Bank of International Cooperation for loans to extend access to new consumers and maintain the present infrastructure. However, the question remains whether such external funding will result in any institutional changes for better management in the light of the past dismal economic performance.

The production, consumption and variable costs incurred by KWA have been analyzed.[5] The fact that a minuscule share is spent for maintenance and the higher share of depreciation reflects the diminishing importance given to the sustainability of the infrastructure created and lowering of institutional capacity, as shown in Table 6.1.

Table 6.1 shows that the installed capacity of KWA has increased substantially over the last 10 years. However, there is a clear

Table 6.1
Production, Distribution and Leakage of Piped Water in Kerala (1999–2006)

Year	Potential Capacity (mld*)	Production (mld)	Distribution (mld)	Unaccounted Water: % to Production
1999	–	1030–1050	770–800	25–30
2000	–	1030–1050	770–800	25–30
2001	1200–1300	NA	1000–1100	20–25
2002	1700–1800	NA	1400–1500	17–22
2003	1720–1820	1260–1360	756–856	37–40
2004	1720–1820	1583.60	1219.37	23
2005	1720–1820	1617.13	1245.78	22.96
2006	1720–1820	1635.00	1259.00	23
2006–07	–	1750	1400	20
2007–08	–	1791	1254	30

Source: Economic Review, Government of Kerala 2006, 2008 and KWA Budgets.
Note: *mld is million litres per day.

discrepancy between production and distribution, showing a huge amount of water (the actual figure is more than 37 percent, although the average reported here is 25 percent)[6] is wasted by leakage that substantially cuts into the efficiency of water distribution. This is also reflected in the wide disparity between the revenue receipts and income (see Figure 6.4). Leakage is a contested term; it is quintessentially unaccounted water. Studies elsewhere have shown that these massive quantities of unaccounted water are actually appropriated by big user consumers like industries and service stations (Bhaduri and Kejriwal 2005).

Figure 6.4 illustrates the wide disparity in the water-selling income of KWA. Though the revenue receipts of KWA are on the rise, the income from water supply has been lesser than total revenue (including plan fund, loan, grant and aid) since 1995–96. The gap moderately widened up to 1997–98, and since then income from water has been lower than total revenue. The water tariffs in Kerala are very low compared to the costs of maintenance of major schemes. The cost per kiloliter of water is INR 8.31 (Economic Review, Government of Kerala, 2008: 140), which is high in comparison to the basic selling price of INR 4 for domestic

Figure 6.4
KWA's Income from Water Supply and Total Revenue Receipts

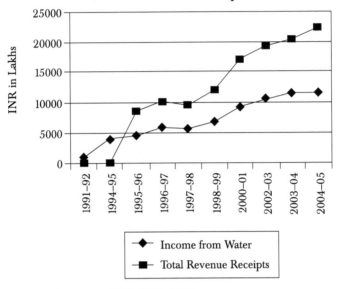

Source: Annual Accounts, KWA (1991–2005).

consumers (see Appendix 5.13 of Economic Review, Government of Kerala, 2008: S66 for details).

Table 6.2 explains KWA's arrears. Local self-governments, including gram panchayats, municipalities and corporations are major defaulters, accounting for almost 77 percent of the total arrears, with domestic consumers accounting for less than 7 percent. In spite of reeling under a financial crunch, KWA is the only para-state body to address the growing demands of, as well as provide free drinking water to, marginalized sections that do not have the ability to pay, through 2,03,874 street taps.[7]

There is a growing supply-demand gap problem, with a steady decline in the per-capita availability of water less than the national average in Kerala.[8] The institutional capacity of the authority in maintaining services to consumers is built by supply-driven state development interventions. Hence the state government support is crucial to carry out the policy of public initiatives to extend access to these resources, which is questionable as indicated in Figure 6.5.

Table 6.2
Category-wise Arrears of Water Charges in Kerala

Category	Arrears Amount (INR millions)	Percentage
Domestic	5509.79	6.98
Non-domestic	9831.97	12.46
Industrial	2611.95	3.31
Gram Panchayat	31036.22	39.33
Municipality	17605.30	22.31
Corporation	12318.67	15.61
Total	78913.90	100

Source: Economic Review, Government of Kerala, 2008: 141.

Figure 6.5
Plan and Non-Plan Expenditure of KWA, 1996–2006

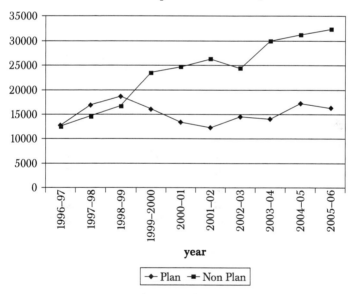

Source: Annual Account, KWA (1996–2006).

The plan fund allocation to KWA from the state government began to decline from 1998, and the subsequent higher rate of increase in non-plan expenditure forced KWA to seek alternate financial sources to meet the necessary investment, including capital expenditure.[9]

Figure 6.6 illustrates the growing disparity between the Government of Kerala grants to KWA and domestic capital expenditure (which includes maintenance and repairing costs). Is this a sign of a state retreat?

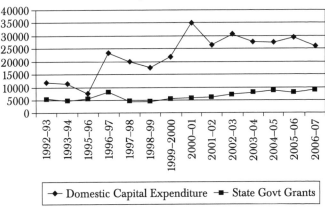

Figure 6.6
Ratio of Government of Kerala Grant to Domestic Capital Investment of KWA

Source: Annual Accounts of KWA.

Foreign Assistance and Institutional Changes

This section describes the inflow of foreign assistance to KWA since its formation in 1984. The institutional changes that this assistance brought about, and the mode by which the assistance was given are analyzed.

Foreign Assistance

Figure 6.7 explains the significance of foreign aid to KWA and compares it with the declining state government funding to it. The World Bank was the first agency which gave KWA aid in 1984. The funding increased till 1994–95, after which it declined steadily and ended in 2000. The Danish International Development Agency (DANIDA) and Netherlands/Dutch government aid

Figure 6.7
Foreign Aid and Government of Kerala Funding for Water Sector

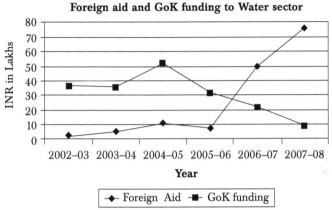

Source: KWA Budget, from 2002–03 to 2007–08.

was the largest bilateral donor to KWA, starting in 1988–89, with the highest share in 1996–97 and ending in 2003. The DANIDA aid continued for 10 years. The Japanese Bank of International Cooperation had started disbursement in 2005–06 and except the Japanese Bank of International Cooperation funds, all foreign assistance was for rural water supply.

A major World Bank-aided project, Jalanidhi was started in 2001–02, but is not listed in KWA's budget documents, since it was implemented through a network of NGOs (support organizations) coordinated by a Government Organized NGO (GONGO) called the Kerala Rural Water and Sanitation Agency (hereafter KRWSA), that works on the principle of a decentralized, demand-driven water supply model, managed entirely by beneficiary groups.

The KWA-led water supply provision has been able to cover 83.07 percent of the urban population and 62.74 percent of the rural population (Kerala Census Report 2001). The greater coverage in urban areas was possible with less transaction costs taking advantage of an already developed urban water infrastructure network. However, due to less rural infrastructure, less coverage was achieved there. The foreign assistance to urban and rural sectors worked on different philosophies and approaches: supply-driven public provision of drinking water to the urban population, and demand driven, community-owned water supply to rural areas – from state to the community, through an emerging water market.

As explained in the previous sections, the KWA was forced to take massive assistance in the form of loans/aid to finance water supply in the state. A major reason for the World Bank loan was the macro policy changes at the national level to shift to more management and result oriented public utilities.[10] Let us examine the modus operandi of two major foreign assisted projects, currently under operation in Kerala for drinking water provision in urban and rural areas.

Japanese Bank of International Cooperation-assisted Urban Water Supply Project

The project is a package of five water supply schemes approved for loan assistance by the Overseas Economic Cooperation Fund (OECF) of Japan, renamed the Japan Bank for International Cooperation in 1996. The objectives of the project are to supplement and rehabilitate the water supply systems of two urban regions and to construct water supply systems for three rural regions, including a cluster of villages. This is the largest water supply project in the state, and when completed, will ensure drinking water to a population of about 4.3 million in the benefited area spread over five districts.[11] It was supposed to commence in 1997 and completed by 2003. However, the appointment of the main consultant for implementation of the project, a mandatory pre-requisite as per the loan agreement, was inordinately delayed and the project started only in 2006.

Of the total cost of INR 17,870 million, the Government of Kerala contributes INR 2,684.5 million and the loan component is INR 15,193.8 million.[12] The project cost was revised to INR 25,899 million in 2007–08. The total amount received as external assistance for this project from 2003–04 to 2007–09 was INR 3,576 million and cumulative expenditure on March 2008 was INR 7156.8 million.[13]

Table 6.3
**Estimated Cost and Loan Part of Japanese Bank of
International Cooperation-aided Drinking Water Project**

Serial No.	Item	Estimated Cost (INR in millions)	Loan Part (INR in millions)
1	2	3	4
1	Scheme 1	2151.2	1994.7
2	Scheme 2	1759.6	1628.7
3	Scheme 3	2313.7	2153.9
4	Scheme 4	4305.9	4102.0
5	Scheme 5	1953.1	1843.7
	Sub total	**12483.5**	**11723.0**
6	Institutional Strengthening	341.8	341.8
7	Contingencies	1282.5	1282.5
8	Consultancy	1000.1	1000.1
9	Interest during Construction	846.4	846.4
10	Land Acquisition	198.6	–
11	Administration Charges	987.6	–
12	Taxes and Duties	734	–
	Grand total	**1787.45**	**1519.38**

Source: Government of Kerala, Economic Survey, 2006.

The project symbolizes Japan's Overseas Development Assistance policy of developing an export market for expertise and loans (Hook 1995: 78). In the Japanese Bank of International Cooperation-funded project, KWA seems to totally depend on the technical expertise of a foreign consulting firm called Tokyo Engineering Company (TEC) for the following – investigation, design, preparation of estimates, assisting in tendering and awarding of works, construction supervision, quality control, including a trial run and commissioning.[14]

The urban water supply schemes in this project are dominantly designed and executed by consultants suggested/appointed by the Japanese Bank of International Cooperation.[15] The main responsibilities of KWA engineers are land acquisition, and award of works, managing the consultants' work, facilitating the contract works, ensuring inter-departmental co ordination and permission from other agencies like the Public Works Department, National Highways Authority of India (NHAI), Indian Railways etc, monitoring work progress, payment to contractors, etc. During the time of fieldwork in 2007–08, there were 34 engineers involved in the project. The organizational structure is shown in Figure 6.8. The consultants have a central office at Thiruvananthpuram, headed by the Team Leader (Chief Construction Engineer) and five regional offices in the scheme areas, each headed by a Resident Engineer. Under the Resident Engineer, there are Site Engineers, one for each contract package supported by senior inspectors and junior inspectors. The consultants manage the project based on a work program agreed with the contractors.[16]

This further raises two points. One, whether KWA engineers, to save consultancy charges as well as avoid dependence on consultants, could have done the design and quality control. Second, whether arguments of transparency and accountability claimed in the consultancy mode of execution is relevant in all contexts. In our initial interactions in 2007, the KWA engineers felt no involvement with or ownership of this new infrastructure creation, since only the administrative staff in KWA was posted in the Japanese Bank of International Cooperation wing.[17] Even though the present structure of the Japanese Bank of International Cooperation shows the apparent involvement of engineers (Figure 6.8), it is unclear whether they were involved in the design phase and also in the field implementation, which raises questions of technological dependency and withering of indigenous expertise. We were unable get details of future technical dependency through service contracts etc, due to the probable lack of capacity in the system.

The significance of the Japanese Bank of International Cooperation funded project is that it is not as transparent as was envisaged. The KWA has not been assigned any role in designing the terms of reference of the international consultancy. There is no mandatory provision on technical know-how transfer in the project.[18] Hence, the KWA may get locked into a long-term dependency with the international consultancy firms, which will eventually weaken the gains of the present infrastructure investment. This is an important point that needs further and deeper investigation.

Kerala Rural Water and Sanitation Agency (Jalanidhi)

The World Bank-assisted Kerala Rural Water Supply and Sanitation Project (Jalanidhi) was conceived in mid-1999. The

Figure 6.8
Organizational Chart of Japanese Bank of International Cooperation-funded Project

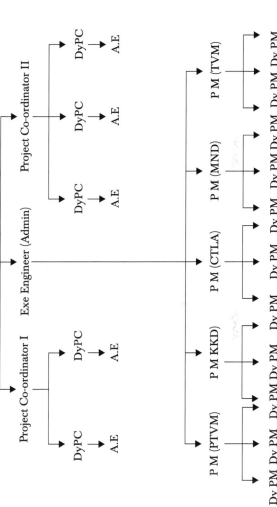

Source: Information received as per RTI Act, 2005 from KWA.

Project Implementation Plan was prepared and project appraisal carried out in mid-2000. The agreement with the World Bank was signed on January 4, 2001.[19] According to Jalanidhi documents, they adopted a demand-responsive approach to service delivery and participatory processes of design, implementation and maintenance. Thus, there is a shift in the role of the government (at the state, district, and gram panchayat levels) from (a) direct service delivery to that of planning, policy formulation and monitoring and evaluation, (b) partial financial support and (c) partial capital cost financing. The total operation and maintenance costs are borne by the beneficiary groups (http://www.jalanidhi.org), which in fact is a policy shift.

Regarding the flow of funds of INR 3,810 million, the Government of India borrowed money for the Government of Kerala from the World Bank (since the Reserve Bank of India [RBI] does not have a mechanism to allow the states to directly receive foreign aid) that is transferred to Jalanidhi and further as shown in Figure 6.4. The implementation is done through an institutional structure totally new to the then structure of water supply delivery in Kerala.

Institutional Sustainability

Jalanidhi is a novel water sector intervention in Kerala that was implemented through a specially created, autonomous, state government-organized NGO (GONGO) and the Kerala Rural Water and Sanitation Agency (KRWSA). The GONGO mobilizes resources and ensures the smooth functioning of support NGOs in community mobilization, without the time and cost overruns of conventional government programs. The KRWSA qualifies as a GONGO because – (a) it is floated by the state government to operationalize the project, (b) it does not suffer from cumbersome government bureaucracy, (c) it can recruit staff in project and consultancy mode, (d) it is supposed to function with flexibility and acquire the management acumen and operational efficiency of an NGO and (e) it has to imbibe the donor-lender, demand-responsive approach with cost-sharing and community participation.

The Jalanidhi project is designed to garner the active participation of NGOs and private consulting firms. According to the Jalanidhi website, they have the following functions and responsibilities – "NGOs or support organizations and private sector consultants will play an important role in the projects. There will be three categories of support organizations. The first will support the Gram Panchayat, beneficiary groups and beneficiary committees on a day-to-day basis in planning and implementing the project's activities, and providing brief support to beneficiary groups post-implementation to stabilize operations. The second is a range of programs for providing specialized assistance, such as preparing groundwater recharge schemes, water quality monitoring, or in developing strategy and materials for sanitation and hygiene promotion programs. The third is private sector consulting firms, which would provide assistance in computerized financial management systems, audits and accounts, design engineering of multi-panchayat water supply schemes, independent construction quality monitoring, management information systems, and monitoring and evaluation in statewide sector development activities."[20]

Jalanidhi invites application from NGOs or support organizations and decides the selection and implementation of schemes. Support organizations will withdraw within three months of completion of the project. Subsequent expenditure and maintenance to keep the project active is the responsibility of the beneficiary groups. The recurring expenditure (including power charge) will be met through the beneficiary contribution. Remuneration of the support organizations has been fixed as INR 3 million per project, which becomes an incentive for the support organizations for initiating Jalanidhi in many panchayats. Once completed, the capacity of beneficiary group members becomes crucial to the sustainability of the project since support organizations withdraw during the completion of the project without any institutional safeguards.

Jalanidhi conducts training for support organizations, including project personnel, with the help of World Bank officials. At the lowest level, the support organizations are directly engaged with the planning and implementation of water supply schemes through a network of 1,519.38 beneficiary groups comprising between

34–55 households. The main objective of training beneficiary groups is to internalize the project approach of cost-sharing and ownership. The mode of state government involvement reflects the nature of GONGO-ization of the schemes with KRWSA entirely facilitating the scheme. The organizational and governing structure reveals institutional changes that KRWSA has brought about in the rural drinking water supply in Kerala. The private on-contract staff working at the implementation level is the major player in Jalanidhi. In the post-implementation period, it seems the entire responsibility of sustainability rests with the beneficiary groups.

Economic Sustainability

The capital cost sharing is done in the ratio of 15 percent by beneficiary groups, 10 percent by gram panchayats and 75 percent by the Jalanidhi project. The entire recurring cost is met by the beneficiaries, which is a departure from the state-run water management system in urban areas. At this point, it is interesting to note that the World Bank appropriated an indigenous model of community-driven water supply scheme planned as part of the decentralized governance experiments in Kerala.[21]

The actual fund flow of the World Bank loan for Jalanidhi is shown in Table 6.4.

Table 6.4
Share of Contributions to Jalanidhi (INR, in millions)

1.	Gram Panchayats' share of 3,227 projects (7.2 percent)	271.6
2.	Beneficiary groups' share (14 percent)	524.4
3.	World Bank through Government of India (72 percent)	2409.1
4.	Other expenses (6.8 percent)	421.1
5.	Total Expenses (as on 30.09.2008)	3626.2

Source: Mid-term review of Jalanidhi: Economic Review, Government of Kerala, 2008: 132.

The future of the scheme is contingent on the collective capacity and economic ability of the beneficiary group members. The sustainability of the schemes is under question if the maintenance expenditure exceeds their capacity.

Table 6.5 illustrates the high operation and maintenance costs incurred by the beneficiary groups (most of them, the rural poor) in a cursory study done to assess their burden. It is seen that beneficiary groups are paying nearly 5 to 6 percent of the average monthly household income for drinking water. This also raises questions about the sustainability of the schemes, with the so-called "demand responsive approach," that rests fully on the financial capacity of the beneficiary group members.

Table 6.5
Average Monthly Expenditure of Beneficiary Groups

Sl No.	Name of Village	No of Beneficiary Groups	Monthly Average Operation and Maintenance Tariff/Household (INR)	Monthly Average Income of Beneficiary Groups	Percentage of Water Expense to Income
1.	Kulathu-puzha	55	45	885	5%
2.	Thalavur	64	52	931	5.5%
3.	Vellarada	42	55	1080	5%
4.	Kulathor	55	67	1100	6%
5.	Mundathikode	62	59	834	7.074341

Source: Average operation and maintenance estimate of beneficiary groups prepared by Jalanidhi as on April 2007, and field survey.

Social and Technological Sustainability

Intervention and Influence at the Level of Ideas

The World Bank consultants and officials have a direct association with the KRWSA. For instance, a sustainability study done earlier by the KRWSA was directly monitored by bank officials.[22] The direct training offered to the project managers of KRWSA and support organizations by World Bank officials indicates the degree of control and influence of the World Bank. It is backed by a strong ideology of a "demand driven approach" in the water sector of Kerala, imported as project philosophy. The discussions center on the binaries of the inefficiency of centrally planned supply driven (KWA) and efficient, decentralized, demand-driven (Jalanidhi) approach. According to the official website, "unlike the supply-driven approach hitherto followed, this project will be based on the needs of the people. The project will be introduced only in areas where interested groups of people show their willingness to participate in the project and abide by the conditions of cost sharing. The group then acquires a legal entity by registering itself and only then proceeds with the rest of the planning. The source selection, technology selection, purchase, contract and implementation is done by this registered body, that is, the beneficiary group, with technical help from support organizations. This includes a sense of ownership in the people."

According to KRWSA's status report (Jalanidhi 2007), there are 170,396 active households in the 4,001 active beneficiary groups, and 3,772 schemes have been completed. The total World Bank loan is INR 357.83 crore. As seen in Table 6.4, the gram panchayat shares 10 percent of costs, beneficiary groups 14 percent and World Bank (KRWSA) 72 percent. The expenditure for maintaining the KRWSA with the foreign consultancy, training and other expenses (including maintaining the central office) is INR 97.46 crore. When we add the core component of beneficiary groups' recurring operation and maintenance costs,[23] the question arises whether the project could have been conceived in an alternative manner without such a huge centralized cost. This is doubly relevant in the light of Kerala's massive decentralization exercise (most vibrant during the Ninth Five-Year Plan period), which empowered the panchayats for the planning and implementation of such projects at the local level.[24] This squarely focuses on the need for local coordination of beneficiary groups at least at the panchayat level, since the localized beneficiary groups might lack the financial and technical expertise to sustain the schemes. The coordination by panchayats is one way to escape the probable institutional vacuum likely to be created by the withdrawal of KRWSA after the project period.

De-skilling of Expertise in the Public Utility

The withering of KWA's technical expertise is a major issue. KWA's seven decades of experience in running and maintaining water supply schemes are currently not utilized in the design or implementation of the projects discussed.[25] In the Japanese Bank of International Cooperation-funded project, the decision to award international and local contracts for design and implementation are exclusively taken by the foreign consultancy organization. This results in over-estimated expertise, bloated expenditure and weakening of the indigenous capacity of the KWA. This might also lead to a perpetual dependency for external expertise.[26]

Cost Recovery Mode

In the rural sector, the modus operandi of the Jalanidhi project raises questions about the criticism of inefficiency of the government system, in comparison with GONGOs. The GONGO model initiated by Jalanidhi is an efficient mechanism, owing to the unparalleled role of the community, through beneficiary groups, in carrying out the schemes with partial capital costs and full recurring expenditure. Hence, the efficiency of the schemes rest primarily on the economic and managerial capacities of the beneficiary group members. However, sustaining the schemes will be difficult, unless the micro level institutions meet the changing demands of operation and maintenance with the availability of water sources.

Operational Efficiency

The Japanese Bank of International Cooperation is unlikely to bring about any managerial innovations in the functioning of KWA, especially with the limited areas of operation, with augmentation of just five urban schemes. The project cell is a parallel structure within KWA with mainly administrative staff, unaware of the technicalities of the project. Considering the lack of transparency, the effectiveness of the proposed capacity-building programs remains a question mark.

Policy Shifts

KWA is a public sector organization based on the conventional model of per capita provision of water connections and public taps, covering rural and urban areas. As mentioned in the third section, there is a total withdrawal of KWA investment from the rural sector. The proposed second phase of the project is supposed to be extended to new areas. Thus, the entire rural water supply will eventually come under Jalanidhi. This is the reason the state government water policy documents categorically made it clear that water policy is "right and resource based," rather than "supply management based."[27]

The continuation of state or supply-driven provision of drinking water to the urban population and policy shift of community-driven, demand-responsive schemes in rural areas is the paradox that is emerging in Kerala. The second model is steered by non-state agencies like a GONGO called KRWSA, which runs Jalanidhi through a network of NGOs and beneficiary groups. The way in which Jalanidhi officials speak, clearly brings out the neoliberal discourse, especially contested terminologies such as "demand responsive approach" and "water as an economic good." These are passed on to the community through the training of beneficiary groups' members and thus carried to the local level.[28] This clarifies that policy change does not always happen through changes in policy documents, but through changes in practices on the ground. In such transformations, the power of ideas is important, as is evident in the discourses within the GONGO-led rural water supply in Kerala.

Consequences of access and control of water for the rural poor

There is a reasonable increase in the access to water in rural areas by these schemes, owing to the participation of beneficiary groups and their willingness to meet the initial expenses and full operation and maintenance costs. As explained in Chapter in this volume, it is as much becoming "modern" through piped water supply as reducing the drudgery. However, the sustainability is dependent on the ability of the beneficiary groups' members to meet any hike in the operation and maintenance costs in future, including repair costs. In the urban sector, as mentioned in Section three there has been no appreciable increase in the access to drinking water. Future studies have to clarify the process of exclusion of sections who cannot "participate" due to their inability to pay as evidenced by Sampat (2007) in Rajasthan.

Sustainability of the Urban Model

There is a serious resource constraint, owing to the dwindling capacity of the reservoir in Thiruvananthpuram and the objection by the Forest Department to expand the reservoir capacity, fearing loss of forest cover. This poses a question as to the perceived benefits of the augmentation plans in the present scheme in the capital city. Demand management by consumers, proper monitoring of consumption by metering and charging penal charges for excess users, supply augmentation by compulsory rainwater harvesting, etc. may be inevitable options in the future.

Conclusion

In Kerala, foreign assistance has played a crucial role in introducing policy and institutional changes in the urban and rural water sectors. However, they work on different philosophies and approaches, such as institution (KWA)-driven, urban water supply through the consultant-oriented Japanese Bank of International Cooperation-funded project, and the community driven rural water sector (Jalanidhi) through World Bank assistance. This signifies a clear state withdrawal from the rural areas and jeopardizes the supply-driven, conventional model in the urban areas, due to the dominance of the consultants. In the rural sector, though the access to water for the poor has increased, it simultaneously puts an additional economic burden on them, with the inflated operation and maintenance cost, repair cost, etc. and the withdrawal of KWA from rural schemes.

Regarding urban schemes, KWA has already extended piped water supply to 83 percent of the population with its indigenous technical capacity, which raises questions about the technological dependency and withering of technical skills of the institutional system, due to foreign assistance. The following issues are relevant in this context – (a) dependence on import of foreign technology, expertise and the absence of technical and institutional capacity building of KWA; (b) financial burden on the public exchequer due to the bloated expenses of the logic of international financing (with consultancies) and (c) low number of additional new connections and thus extending access to drinking water.

The complete relegation of rural water services to beneficiary groups is unsustainable technologically, economically, socially and environmentally, with respect to local sources. Technical and institutional hand-holding by locally based NGOs or any civic organizations, including linking beneficiary groups to the gram panchayats could be another role of civil society at the operational level. At the strategic level, a network of such local NGOs/CBOs at the state level could facilitate policy advocacy to sustain these efforts.

Even without the issue of foreign funding, there are serious problems facing the public utility of KWA. The state government's approach towards KWA, with the declining plan support and appropriate mechanisms to monitor costs, has led to the financial crisis. The state, with its current mode of functioning cannot provide water services to everyone. The effectiveness of public utilities like KWA has to increase with a better work culture, accountability, transparency and community participation. The modalities, roles and levels of public-private (or more rightly public-public) participation by an array of civil society organizations (voluntary, NGOs, citizen/neighborhood groups and other local organizations) with the state has to be specified. There should be informed public debate on water governance-related issues like the demand, supply, costs, tariffs, cross-subsidy for the poor and state support to make the water services efficient, equitable and sustainable, ensuring which is one of the major tasks of civil society organizations.

At the level of ideas, the "demand-responsive approach," "water as an economic good," and resulting "potential water business" generated by international financial institutions, have to be re-examined, debated and contextually analyzed, with an exploration of the alternatives. The study also clarified that such changes in policies do not always happen through alteration of policy documents, but through changes in practices on the ground. In such transformations, the power of ideas is important and the development of alternative ideas could be another valuable role that civil society can play in democratizing water governance.

Notes

1. Pillai, Velu, 1946, Vol IV, chapter 6.
2. Chapters 6 and 23 of the Sixth Five-Year Plan document of the Planning Commission state that "in the States, the Electricity Boards are incurring huge losses. In the case of irrigation, the gross receipts are not sufficient to cover even the working expenses. Most of the State Road Transport Corporations are also making losses. It would be necessary for these undertakings to improve their financial performance through a revision of tariffs, water charges and taxes and other suitable measures" (http://www.planningcommissionofIndia.nic.in). This declaration helped the state government to turn the Public Health Engineering Department to a quasi-state agency and also accept the World Bank aid in 1986.
3. News Vol 1, No. 2, Association of public health engineer of Kerala, September 9, 2005: 9.
4. Interview with the AITUC-led Engineers Union and trade union leaders of KWA. They also admitted that data regarding the infrastructure details are not available to the public.
5. KWA produces 596775 Ml/Y and sells 459535 Ml/Y. The total cost is INR 339.51 crore per year and direct variable cost is INR 169.90 crore, which includes power charges (INR 108.00 crore), indirect overhead cost (INR.169.6 crore) and administrative costs (INR 78.11 crore). The depreciation cost is 34.57 percent and the maintenance cost is 7.27 percent. (Economic Review, Government of Kerala 2007).
6. KWA engineers reported this fact repeatedly to us. They also pointed out that the huge maintenance costs needed to be reduced, along with the leakage. They justify the wastage with the reason that Kerala has enough supplies of water, which by controlling, is questionable in the future.
7. Economic Review, Government of Kerala 2008: Appendix 5.12: S 65.
8. Around 17.2 percent of the villages in Kerala do not get any benefit of protected water supply. Out of the 9,776 rural habitations, 228 are not covered and 69 percent of villages partially covered. In partially covered areas, 14 percent get only less than 10 liters per day and 55 percent get less than 20 liters per day (CWRDM 2005).
9. More than 37 percent is utilized for salary, 33.33 percent for power, operation and maintenance (9.94 percent) and the rest for payment and repayment of loans (Annual Accounts of KWA 2005).
10. Chapter 6 and 23 of Sixth Five-Year Plan of India.
11. Economic Review, Government of Kerala, 2006.
12. Infrastructure investments costs INR 1,248 cores and consultancy and administrative expenditure cost INR 539.1 lakh, of which consultancy alone constituted INR 100.01 crore which is the highest individual expenditure. In infrastructure expenditure, two urban centers

Thiruvananthapuram and Kozhikode costs INR 215.12 crore and INR 430.59 crore, respectively. Other three components of the infrastructure expenditure is maintenance of rural water supply schemes in three panchayats—Meenad, Cherthala and Pattuvam; which costs INR 175.96 crore 231.37 and INR 195.31 crore, respectively. All the administrative expenditure is being met by a loan (Economic Review, Government of Kerala, 2006: Chapter 5).

13. Economic Review, Government of Kerala, 2008: 131.
14. Written reply from the Managing Director of KWA to the queries raised by us through the RTI Act 2005.
15. In the Japanese Bank of International Cooperation project, there are 28 contract packages of which 11 are international competitive bidding and 17 local competitive bidding (Economic Review 2006). A controversy erupted in July 2007 about the quality of pipes imported from China worth INR 2 billion with the minister conducting a press conference. All the Malayalam dailies (see *Malayala Manorama* and *Madhyamam* published on July 12, 2007) reported this press conference.
16. Information received as per RTI Act, 2005 from KWA. (No KWA/JBIC/CE/3977/07) dated August 16, 2008.
17. Interview with Assistant Executive Engineer P.S Rajeev in KWA on August 9, 2007.
18. Letter in response to request under RTI-Act 2005, (No KWA/JBIC/HQ/E 8036/07) dated March 25, 2008.
19. Government of Kerala, Economic Review 2002.
20. The details of the responsibility are as follows: (a) Conduct base line survey of the panchayat – the socio economic status of village population where the scheme is proposed. This survey is the base on which the SO fixes the nature and cost of the schemes. (b) Introduce the project in the gram sabha. This includes conducting introductory rally through the villages and attending the gram sabha along with the panchayats members. (c) Conduct training for villagers regarding the nature and philosophy of the project. It is crucial to scheme. The prime concern of this training is to convince people about the methodology of demand driven aspects of the project. (d) Organize beneficiary groups based on the economic capacity of the households. (e) Select the appropriate project including project cost and BG's share. (f) Train the beneficiary groups – b1. Select the pump operator (beneficiary group ember) and impart training to operate; 2. Conduct training on water purification – chlorination; 3. Help the beneficiary groups to select an appropriate source of water (well); 4. Help purchase the accessories for instance, pipes; 5. Select land for constructing the well and assess the water carrying capacity of the well; 6. Supervision of construction. The role of SO in

the post-implementation periods are; (a) Calculate the O&M based on monthly power charge, pump operator's salary, and (b) Fix the monthly O&M per family in BG (http://www.jalanidhi.org).
21. See the experience of Olavanna Panchayat by Elamon, Joy 'People's Initiative In Water – Olavanna Village In Kerala, India"; Reclaiming Public Water Shows The Way: Achievements, Struggles And Visions From Around The World: Page 44. TNI
22. Jalanidhi, 2007, Report on Jalanidhi (Kerala Rural Water Supply & Sanitation Agency): Sustainability Evaluation Exercise-IV: Thrissur and Palakkad districts. Thiruvananthapuram: KRWSA.
23. The primary survey of Kulathoor Panchayat of Thiruvananthapuram district alone shows that nearly 15 percent of beneficiary groups have paid for additional maintenance. The share of beneficiary groups' contribution and panchayat come around INR 71.95 crore. The average monthly operations and maintenance cost is INR 52/households. The local operations and monitoring cost of the project is INR 12.48 crore, excluding the maintenance cost.
24. The interesting point is that even the World Bank's Jalanidhi model rests on one such model from the Olavanna panchayat in Northern Kerala (Elamon Joy, op cit).
25. The Assistant Engineers Association of KWA made many memoranda to the state government explaining the history and technical efficiency of KWA in implementing drinking water supply schemes but none of them were noticed.
26. In the reply to our queries through RTI Act, the Japanese Bank of International Cooperation cell has clarified that "...the consultant's international experience will pervade the design of the various components of the scheme" Japanese Bank of International Cooperation Assisted Kerala Water Supply Project – Information Furnished Under Right to Information Act 2005.
27. Draft of Kerala State Water Policy, Government of Kerala, 2006: 32.
28. The Jalanidhi officials tell the beneficiary group members that this is the only way out for them. We have witnessed the same in our field work.

References

Bhaduri, Amit and Arvind Kejriwal. 2005. "Urban Water Supply: Reforming the Reformers," *Economic and Political Weekly*, 40(53): 5543–45.
Centre for Water Resources Development and Management (CWRDM). 2005. "Water Resources of Kerala." Kozhikode: CWRDM.
Dharmadhikary, Shripad and Gaurav Dwivedi. 2009. "Water Sector Reforms and their Implications in Madhya Pradesh'. Paper presented

the International Conference on Water Resources Policy in South Asia December 17–20, 2008 Colombo, Sri Lanka. Organised by South Asia Consortium for Interdisciplinary Water Resources Studies (SaciWATERs), Mimeo.

Government of Kerala (GoK). 2006. Draft of Kerala State Water Policy 2006.

———. 2006. *Economic Survey*.

Cullet, Philippe. 2006. "Water Law Reforms: Analysis of Recent Developments". New Delhi: International Environmental Law Research Center.

Elamon, Joy. 2005. "People's Initiative in Water – Olavanna Village In Kerala, India, Shows the Way," Transnational Institute (TNI).

Gorringe, P. 1997. "The State and Institutions," The Treasury, Wellington New Zealand. Available online at http://www.treasury.govt.nz/gorringe/papers/gp-1997.pdf (accessed on February 10, 2004).

Jessop, B. 1994. "Post-Fordism and the State," in A. Amin (ed), *Post-Fordism: A Reader*. Oxford: Blackwell Publishing.

Jalanidhi. 2007. Report on Jalanidhi (Kerala Rural Water Supply & Sanitation Agency): Sustainability Evaluation Exercise–IV: Thrissur and Palakkad districts. Trivandrum: KRWSA.

Kerala Water Authority (KWA). Account documents since 1985. Government of Kerala: KWA.

———. Budget Estimates, various years. KWA.

Mackintosh, M. 1992. "Introduction," in M. Wuyts, M. Mackintosh and T. Hewitt (eds) *Development Policy and Public Action*, pp. 1–12. Oxford: Oxford University Press.

Association of Public Health Engineer of Kerala. 2005. "News," 1(2): 9.

Pillai, Velu. 1946. "The Travancore State Manual," Vol IV, Kerala Gazetteers Department, Thiruvananthapuram, 1996 (reprinted in 1996).

Rhodes, R. A. W. 1997. *Understanding Governance. Policy Networks, Governance, Reflexivity and Accountability*. Buckingham and Philadelphia: Open University Press.

Sangameswaran, Priya. 2010. "Rural Drinking Water Reforms in Maharashtra: The Role of Neoliberalism", *EPW Economic and Political Weekly*, 45(4): 62–69.

Wagle, S. and K. Dixit. 2006. "Revisiting Good Governance: Asserting Citizens Participation and Politics in Public Services," in D. Chavez (ed.), *Beyond the Market: The Future of Public Services*, pp. 21–30. Amsterdam: TNI/PSIRU.

PART III
NGOs in Policy Influence

7

Water Policy Development in a Multi-party System of Governance

A Case Study of Sri Lanka[1]

Rajindra Ariyabandu

Water Policy Development: A Continuing Process

For more than a decade, Sri Lanka has been struggling to formulate a comprehensive water policy for management of water resources. Although cursory attempts were made in the 1980s, no concerted efforts were made to institutionalize water resource management. However, since the early 1990s, two distinct projects, the Comprehensive Water Resource Management Project (CWRMP)[2] and Water Resource Management Project (WRMP)[3], both supported by the Asian Development Bank (ADB), attempted to streamline policies, legislation and institutional arrangements to achieve more efficient and holistic water resource management. However, they failed due to poor project design and lack of awareness of the country's context, in particular the cultural and political nature of water.

Sri Lanka's approach to water resource management has focused mainly on alleviating seasonal water scarcity through construction of reservoirs and trans-basin diversions, with their roots in ancient hydraulics. Sri Lanka has a long tradition of water management. Against this backdrop, many donors have played a key role in water resources management in Sri Lanka. The most recent water policy process supported by ADB, had its predecessors in USAID and

World Bank-projects. These agencies have been key actors in the implementation of several important water resource management projects over several decades.

In the 1970s and 1980s, USAID and World Bank-supported projects initiated policy development aimed at enhancing productivity in the irrigation sector. These projects promoted farmer participation and introduced charging fees for irrigation water. These moves were seen as the commodification of agricultural water, and many farmers withdrew from participating in these projects. These interventions did not go beyond the pilot stage, mainly due to poor participation (Bandaragoda 2005). Learning from this experience, the Gal Oya Water Management Project, supported by the USAID, took a significant step in introducing "farmer participation" in irrigation water management, and introduced the concept of "institutional organizers" to facilitate this. At that time, many donors were actively supporting the construction of large dams and other infrastructure for water management. However, the dam boom which dominated the 1980s, gave way to greater emphasis on sector management by policy reforms and financial control.

At the same time, the failure of many irrigation projects led professionals and policy makers to promote institutional reforms in the irrigation sector. The reforms suggested included, new policies in the irrigation sector, legislation and emphasis on basin level organizations. The Irrigation Management Policy Support Activity (IMPSA), a USAID/ISPAN (Irrigation Support Project for Asia and the Near East) project, supported policy development in the irrigated agriculture sector in the early 1990s.

The 1990s was a decade of water sector reforms in South Asia. Many countries attempted sector reforms aimed at providing better service to water users, especially the poor. These reforms were driven by a mixture of internal factors like water scarcity, performance of the water sector deterioration and financial non-viability, and external factors like macro-economic changes, political reform and institutional arrangements (Saleth and Dinar 2004*). In Sri Lanka, external factors were the drivers for reform. Although water scarcity exists in the dry zone, internal strategies typically focused on infrastructure development to counter the effects of droughts and dry spells. Policy and institutional reforms were advocated by the donors.

The IMPSA project recommended the establishment of farmers' organizations for irrigation system management, and advocated system take-over beyond tertiary canals. It also recommended that farmers be exempted from operation and management (O&M) fees in lieu of system take-over. These recommendations were viewed by some as the failure of the state to protect and support farmers. Some even feared that these were the first steps towards commodification of water, and the involvement of the International Water Management Institution (IWMI) was part of a conspiracy to sell water resources in Sri Lanka.[4] This actually implies privatization of water resources leading to a change in ownership of water resources which amounts to selling of water.

The changes in Sri Lanka's perception of water resources management coincided with changes in the global discourse on the subject, from dams and development, to policies, institutions and management. This shift was evident in Sri Lanka when some donors advocated the adoption of "water rights" to maximize efficiency and productivity of water by moving away from high water-consuming crops to high-value, low water-consuming crops. This was prescribed in the 1996 report on non-plantation crop sector policy alternatives for Sri Lanka. Concurrently, the Dublin Principles[5] emphasized the importance of water as an "economic good" and the need for countries to adopt Integrated Water Resources Management (IWRM) through active stakeholder participation as the future strategy for holistic water resource management. Also, the global consensus on policy and institutional reforms for effective water governance coincided with the strategy on water governance and sustainable economic development in South Asia. Thus, the global and regional initiatives resulted in water sector reforms in Sri Lanka. To maintain policy development in an apolitical environment, neutral ministries were chosen to take the responsibility for the policy development process. Sri Lanka has a large number of ministries dealing with the subject of water. Water policy formulation vested with any one ministry dealing with water, could result in bias towards that ministry. Hence, it was envisaged that a "neutral ministry", which does not have any links to water development, would perform a non-partisan function in policy formulation. Hence, the Ministry of Finance and Planning (MF&P) was given the responsibility of driving the process.

Following these initiatives and IMPSA recommendations, the Sri Lankan government made a proposal to ADB to develop a "Water Resources Master Plan" in 1992. This led to the CWRMP, which was to drive the policy development process over the next decade. The guiding principles of CWRMP were holistic development and efficiency. Once again, the emphasis on "efficiency" led to concerns in civil society that the project was an attempt to commodify water. In the mid-1990s, the CWRMP was supported by the Food and Agriculture Organization (FAO), Netherlands, which assisted the Water Law Policy Advisory Program draft the Water Resources Act.

The CWRM projects and their related activities focused on a single overarching apex body which was to govern the water resources sector. With over 50 laws and some 40 institutions dealing with the subject of water, the reform process (through its apex body) was to streamline water resources management and sector development processes. From the beginning, the CWRM project had a relatively uninterrupted passage because its "ownership" firmly rested with the Ministry of Finance and Planning. This meant minimum conflict of interest with the policy development process. In 1996, following some resistance to the proposed reforms, the Sri Lankan government established the Water Resources Council (WRC) and the Water Resources Secretariat (WRS). The WRC was a high-powered advisory committee comprising Secretaries to ministries involved in water development, line agency heads, non-governmental organization (NGO) representatives, academics and representatives of farmers' organizations. The WRS consisted of "seconded" technical experts, who were responsible for the formulation of the policy and legislation for government approval. The final challenge for the WRC and WRS was to establish the National Water Resources Authority (NWRA) as the water "apex body." The WRS carried out stakeholder consultations during 1996–2000, which led to the development of the "National Water Resources Policy and Institutional Arrangement." This was approved by the Cabinet of Ministers in March 2000. The Water Resources Act, which was to follow the water resources policy, would formally establish the NWRA, gave legal recognition to the River Basin Committees and establish "water rights." However, due to an intense political debate and resistance, this never came about and the Act was never presented before the Parliament.

The absence of a law enabling management of water resources had different implications for various water stakeholders. The NGOs and civil society organizations (CSOs) who opposed the policy presumed that they had achieved their task. Large industries continued unabated extraction of groundwater, depleting resources and affecting the environment. Water allocation conflicts between irrigation and drinking water continued, with some conflicts being arbitrated by the court of law. Small water users in irrigation systems had to struggle with the authorities to get their share of water. On the other hand, "potential water entrepreneurs" who anticipated the passage of the law in Parliament, were disappointed that they could no longer have access to "water entitlements" for potential water trade.

Water Resource Management Project: The Final Intervention

In 2001, ADB funded another project, the Water Resource Management Project (WRMP), primarily to strengthen the NWRA and related organizations, and the construction of much needed infrastructure in selected river basins to improve the quality and quantity of water. The two components of the project, Part A, water resource management and institutional strengthening, and Part B, construction of the Kaleni Conservation Barrage, were kept separate, managed by two different institutions.[6] Although the two parts had different objectives, they were linked together as one project with the major covenant being the establishment of NWRA.

However, the rationale for linking the two projects was not clear even when the project was completed in 2006.[7] Continued opposition to the policy and the institutional arrangement, led to premature closure of the project. The WRS attempted to drive the process, but without any legal authority, it could make limited progress. In 2004, Part A of the project was suspended, and the infrastructure building component, that is, Part B had to be suspended in favor of alternatives proposed by the irrigation department in 2006.

The bank accepted many requests from the government for time extensions to meet project covenants, and supported a consultation and awareness-building process (Part C).[8] However, the proposed consultation under Part C did not materialize due to inaction by the government and non-compliance with the major covenant of the project.

As a result of these impediments, key institutional reforms could not be achieved and the capacity building component of WRMP remained in jeopardy. The project suffered from political resistance, changes in the government and mistrust of civil society, who argued that consultation and communication on the new policy was inadequate. They also feared that approval of the policy would mean commodification and privatization of water resources in Sri Lanka.

Resource Availability and Demand Management

With a per capita water resources availability of 2,400m^3 and annual average rainfall of 2,000mm, Sri Lanka does not have an immediate problem of water scarcity. The 103 distinct river basins in the country discharge nearly 35 percent of the annual run-off into the sea. Around 20 rivers in the wet zone carry almost half of the total run-off. Although ground water assessments are weak, it is estimated that potential groundwater availability is approximately 7,800 million cubic meters (mcm). However, the quality and quantity of ground water limits its usage for productive purposes (Imbulana et al, 2006*).

As a result of very high spatial and temporal variations, some districts of Sri Lanka experience prolonged dry spells or droughts. As an island, Sri Lanka receives all its water from precipitation. Out of 43,000mcm of renewable water resources, only 11,000 mcm can be used for productive purposes. Over 80 percent of developed water resources is used for irrigation, while only a fraction is used for domestic and industrial purposes. It is assumed that approximately 80 percent of rural drinking water comes from groundwater, while surface water provides for the majority of urban water use. However, there is a lack of reliable data on water use; different studies have given different estimates.[9] While surface water flow data are available with the Irrigation Department and National Water Supply and Drainage Board, the reliability of groundwater data is rare. This lack of reliability has led to the promotion of different approaches to water resources management, rather than a cohesive strategy for development.[10]

It is commonly believed that there was ample water for irrigation before economic liberalization in 1977. The demand for irrigation water was less, and introducing IWRM was not a pressing issue (ADB 2000). However, the situation has changed rapidly, with industrialization and urbanization[11] demanding greater and more reliable provision of safe water. With current estimates of 8–10 percent annual increase in demand for safe water for domestic use and 100 percent domestic water coverage target by 2025, the water supply department will have to look for alternative sources, like rainwater harvesting (Wickramage 2002).

According to current law, groundwater is tied to land rights. Anyone with legal rights to the land has the right to access unlimited quantities of groundwater. This has resulted in urban water supply schemes and industries extracting water without control or consideration for the recharge capacity. Besides, any water sources conveniently accessible have been used as an outlet to discharge waste water, jeopardizing both surface and groundwater quality. The current practice of excessive sand mining has altered river morphology, causing severe stress on groundwater, and salt water intrusion into surface water bodies.

The inefficient use of water resources has been exacerbated by the fact that the resource has been exploited by a variety of agencies for their own gains. Some large water users, including government agencies, allocate water to themselves, thus playing the role of both regulator and user of the resource. Although they may take other users to account, they have no legal obligation to allocate water in an impartial manner. This is evident in the current practice of water allocation of the Kelani river.[12] Further, water requirement for the environment has not been protected since there was no recognition for such use prior to the recent policy debate. Hence, there was no mechanism to maintain minimum in-stream flow or reservoir levels for environmental needs. Besides these issues, urbanization has led to competition between irrigation and drinking water supply,[13] and irrigation and hydro power. Thus, there is a clear role for demand management approaches for equitable distribution of water. However, in the recent past, demand management has been widely equated with charging fees for water use. Indigenously designed measures such as the construction of small reservoirs which save water, benefit livelihoods and minimize the impact of spatial and temporal variations, have been largely ignored by the state in favor of pricing instruments.

Policy Development Process

Controversies and Assumptions

Sri Lanka has many ways of establishing policy—by the constitution, governing party manifesto, public investment programs, the annual budget and the policies adopted by the Cabinet (Mosley 1994*). In all these cases, the water resource policy has been guided by the political forces in power. As indicated by policy preparatory technical assistance (TAs), there was no consistent policy for WRM. This could be due to lack of reliable data and information to assess water use and scarcity. At the same time, the demand for available water was less than supply, and there was no apparent competition among water users. Thus, there was little perceived need for high level water resource planning.

However, the 1980s and 1990s saw increasing consensus on the need for a new comprehensive WRM policy. The increase in demand was a result of competition among water users and multiplicity of institutions dealing with water resources. Some even alleged that rational projects and programs could not be implemented due to lack of a clear WRM policy (Dharmasena 2005). However, many policies and policy development processes that were proposed, lacked the Sri Lankan context, foresight and awareness about politics, media and the public. This weakness later developed into much controversy surrounding the policy development process. Two clear examples in this regard are the establishment of the "apex body," NWRA, for all water development and management functions. This created a lot of controversy and generated considerable debate, confusion and opposition. Proponents of the apex body concept believed that the integration of water sector institutions under one umbrella could solve many of the current problems faced by water resource management. Those who opposed, protested that many traditional water institutions could be threatened with the concepts of the apex body, that it would only add to the already overburdened state institutions, create pressure on the national treasury and make the implementation of legislation more difficult.

The second issue was the introduction of "bulk water use" which created great uncertainty among water users and professionals alike. The policy which introduced the concept did not offer a clear definition for "bulk water users." The definition later given in the draft law was weak and ambiguous. This ambiguity led to the speculation that the policy was favorable for bulk water users, at the expense of small water users, who were in a majority. Some even extended the argument to cover farmers' organizations in irrigation systems, and that they should be charged for water use. The farmers' organizations were expected to pass the financial burden to individual farmers, which eventually ended in charging poor and marginal farmers for irrigation water use.

However, the water resource law, which was drafted many years ago, has still not being approved by the Parliament of Sri Lanka. While there was consensus on the need for a policy of water allocation and management, there was no agreement on how this would be determined. Some donor agencies believed that rational allocation of water should be promoted by establishing a system of "tradable water entitlements" designed to discourage high water-use crop (e.g. paddy). Water would naturally move towards high-value uses, as trade in entitlement occurs within and between sectors at freely negotiated prices. Proponents of this theory expected rural towns to pay farmers for release of their water allotment, and hydro power and industry were expected to expand rural employment, as their water supply would be secured through formal entitlements. Unfortunately this approach was based on the assumption that Sri Lanka suffers from serious water shortages resulting mainly from misallocation of water.

These proposals clearly indicate lack of knowledge of local context. Paddy cultivation is not only a source of food security, but a cultural tradition. There are many reasons for this. It imparts a sense of land ownership, even with a small plot size. Hence, such approaches are unlikely to change the pattern of cultivation or encourage a shift away from paddy cultivation. Although the need for a new policy was accepted in principle, the suggestion to move away from paddy cultivation angered the public and reduced the credibility of donor initiatives. This was reinforced by global events in which the poor had to sacrifice their water rights to multi-nationals, as in the case of Cochabamba in Bolivia. Hence, the water policy process was begun in a climate of controversy and suspicion, which set the stage for events up to the present day.

Policy Formulation Process

Initial Formulation

Since the inception of the policy development process, the formulation of policy was the responsibility of the Senior Technical Advisor (STA) to the ADB, supported by a team of local professionals seconded from the relevant government institutions. Although the local counterpart team was expected to be multi-disciplinary, it never functioned as one. Crucially, it lacked the services of a sociologist and an economist on a long-term basis. This meant that the focus was often more technical and key socio-economic issues failed to get strategic attention. Besides, team members' involvement in the process was not consistent, due to administrative and ideological differences on crucial policy issues.

The policy process and the new institutional arrangement were to be housed under a neural ministry, with ownership extending beyond the policy development process and hosting NWRA. However, concerns were raised by some WRC members as to the effectiveness and capacity of a neural ministry to host the "ownership" of policy[14] implementation. Thus, there was no clear consensus even among the WRC members on the future status of NWRA, which was supposed to take the water sector reforms forward.

As a result, the WRC agreed to propose that NWRA be housed under the President, the highest office in the country. Although most WRC members agreed on the "compromise," some believed that the President should not be burdened with "unpopular decisions." This clearly indicated the uncertainty among the WRC members who were expected to function as the advisory council to the NWRA after its formal establishment.

The policy development process was over-burdened, with mundane administrative problems overriding some of the key policy issues. The turnover of local counterpart staff to assist the STA was a constraint and staff support from the irrigation department (a key counterpart organization) was intermittent. With all the constraints to policy development, "policy ownership" was a serious concern. Over an eight year period, chairmanship of WRC changed nine times and the consequent loss of institutional knowledge seriously affected the functioning of WRC and the policy process. Although changes in political administration were cited as the main cause for high turnover of the chairmanship, certain undercurrents within the bureaucracy too may have contributed to the quick turnover. As a result of all these, the WRC postponed submitting the policy proposals to the Cabinet, at least twice, stating the time was "inappropriate."

Stakeholder Consultations

The policy consultation process took place through consultation and meetings, mainly among water professionals. An attempt to consult small water users, mainly farmers through the involvement of an NGO was not very successful. Although it was not evident at the time of policy formulation, it was later revealed that poor consultation with small water users was due to an internal rivalry between the NGO selected by the Water Resources Secretariat (WRS) for the purpose and other field-based NGOs and CSOs. However, it was estimated that the WRS conducted around 115 consultations with the assistance of local and foreign consultants. Besides, formal consultations, WRS consulted a number of research and policy institutions on the content of the policy.

The policy consultations were often not very comprehensive and transparent. Awareness raising issues were targeted at large audiences, while contentious issues were discussed with a selected few. Inconsistent local support staff and lack of understanding on policy content could be blamed for segregated policy consultations. Key policy documents, including the "National Water Resources Policy and Institutional Arrangement" were published in English, which made it inaccessible to the majority of the population.

Besides, only 300 copies of the approved policy were published, making it available only to a selected few. In some of the public information brochures, sensitive issues like entitlements, bulk water use, tradable water rights and cost recovery were watered down to avoid public disapproval. In the process of consultation, Cabinet of Ministers too were reluctant to lend a formal backing to the process, fearing public disapproval and their own political careers.

The WRS/WRC therefore, decided to take a different approach and published an information brochure on water sector reforms. Thereafter, the political aspects of water reforms were removed, fearing public protest. In the end, the information brochure discussed some non-controversial IWRM issues. The WRS/WRC also established a coordination committee with water research institutions, but the consultation with these institutions too did not yield any significant outcome. This, again, was a good attempt without any tangible results due to poor strategy, commitment and understanding. Thus, the weakness in the consultation and communication process angered the public and reinforced suspicions about the content of the policy proposals.

Despite the controversy surrounding the proposals, the policy was approved by the Cabinet of Ministers in March 2000. This was hailed as a landmark achievement by many donor agencies at the time. The policy process and the new institutional arrangement remained within the Ministry of Finance and Planning. This was an advantage in terms of gaining Cabinet approval as the Minister of Finance and Planning at the time was the President of Sri Lanka. However, confusion continued when the "draft act" had to be approved by the Parliament. The legal administration determined that the consultation process should be completed within a month, before the act was submitted to the Parliament.[15] The undue haste made the public suspicious about the contents of the reform proposals. Opposition to the proposals was mainly based on distorted facts. Allegations were made that reservoirs will be sold to multinational companies and people will have to pay for water from all sources, including their own household wells. The WRS/WRC, in an attempt to counter these allegations, decided to publish the policy guidelines in newspapers for public information. However, this could not be accomplished due to undisclosed reasons.[16] This, again, reflects lack of a clear strategy for public consultation and political reluctance to accept ownership of the policy formulation process.

Advocacy or Pressure Groups: Role of Civil Society

As a result of the lack of public information and understanding of issues of public policy, and limited experience, civil society came to play an active role in the water policy development process. Usually, civil society, represented by NGOs, plays an important role in raising awareness about environmental concerns and promoting sustainable development which is why the policy makers approached NGOs as contributing partners in the water sector reforms process. Civil society was represented by the various environmental NGOs who were essentially concerned about protecting the water resources of Sri Lanka. While there was no debate about protecting natural resources for the livelihood of present and future generations, the approach adopted by some NGOs left much to be desired. The complexity of the state water policy attracted severe criticism and led to large scale misrepresentation. The fundamentals of the policy were based on fixing a price on water. When water is a commodity, its use will be decided by market forces. Whether this was the correct approach for equitable water distribution remains a question that is unanswered till date. The hierarchy of the engineering fraternity governing the reform process could not explain the reason for establishing tradable water rights. This gave vent to various interpretations by interested parties, including the CSOs.

Civil society, when asked to explain the "unexplainable," took a radical approach. They simplified the complexity of the policy by providing vivid descriptions of the "aftermath" if the proposed policy was passed by the Parliament. The strong environment NGO lobby was fully backed by the media, and some of the political parties who were opposed to incumbent regimes. There were three civil society groups protesting against the water policy–those sincerely concerned about environmental implications and the livelihoods of the poor, those who wanted to expose the global water "mafia" and the nexus between local and global water traders, and those who wanted to derive political mileage from the debate. All the three groups made use of the media to raise their concerns, while the media itself used the issue to its own advantage in isolating and attacking the political leadership at different times in the policy process.

In a multi-party political system, where the survival of political regimes depends on the strength of the coalition partners, subsequent governments were unable or unwilling to defend many of the contentious issues raised by civil society against the policy. The NGOs bulldozed their way through intense protests, leaving no space for professionals to explain the benefits and disadvantages of the policy. The media also intensified the campaign against the policy through multiple media, to good effect. The message, by civil society, media and politicians was the same—water is an economic good and it will carry a price in future. Some media feared the wholesale of large reservoirs to multinational companies.

Looking back, we see that some beneficial aspects of the policy could never have been implemented. In the current context of climate change and increasing natural hazards, water resources are becoming scarce. In the future, only those with power will have access to water, leaving the poor and marginalized at the mercy of the rich and the powerful. It is the duty of the state to protect the poor and the marginalized, maintain equity and improve livelihoods. Given the dubious history of the policy, civil society was expected to identify gaps and challenges in the water policy, inform the public, and work towards sustainable water resources. However, today there is more confusion and increased conflicts regarding access to water.

Water Resource Management Project: Wrong Presumption in the Right Direction

The WRMP was the natural extension to earlier efforts to institutionalize integrated water resource management. The CWRM project had identified several institutional gaps for effective water resource management in the country. The role of the WRMP, therefore, was to identify these gaps and strengthen the institutional arrangement for effective WRMP. At the time of project preparation, the project was consistent with the government policy for establishing adequate mechanisms to ensure sustainable water resource management in the country. Thus there was common agreement that the country should be moving towards efficient and holistic management of water resources. The need for an effective national agency to operationalize the water policy and oversee the management of the resources was considered mandatory. Hence, the government and ADB gave priority to establishing the NWRA, as it was seen as a natural development arising from work carried out in earlier projects through the CWRM.

The project design was also consistent with the ADB policy on integration of water service delivery and water resource management in its member countries (ADB 2007). Further, it was also aligned with the ADB country strategy that "support for comprehensive water resources management would consolidate recent institutional changes in the sector, strengthen the capability of the recently established Water Resource Council and Secretariat as the water apex body, and facilitate agency restructuring in the water sector, leading to the formation of an agency responsible for water resource allocation and management." Thus, the WRMP, particularly Part-A[17] was expected to be sound and coherent, given the initial exploratory work and international experience.

Although the project design was relevant and genuinely attempted to establish sustainable WRMP, its conceptualization and implementation were faulty. For instance:

(a) The inclusion of the establishment of NWRA as a major covenant and also as a pre-requisite for a project which contains financing of a physical infrastructure (Kaleni conservation barrage), can be justified without the establishment of the NWRA.

(b) The inability to anticipate stakeholder resistance or lack of acceptance for a comprehensive water resource management policy and an institutional arrangement.

(c) Lack of understanding on the perceived threat to existing water sector institutions by creating a new powerful apex body.

(d) Poor knowledge of perceived notions of water privatization through sector reforms.
(e) Linking Part-A and Part-B[18] of the project which were independent and implemented by two different executing agencies.
(f) In the absence of a permanent arrangement for WRMP, the project deviated from the original goal of strengthening the institutional arrangement, and entered into a political process of policy and legislation approval by Parliament as a time-bound covenant of the project.

With mounting donor pressure for policy approval, WRS made several ad hoc changes to the policy. As a result of this, the final draft which was published for public comments in October 2006,[19] carried many contradictions.[20] In the absence of clear "ownership" to the policy development process, WRS assumed a "de facto" ownership, but it did not have the capacity or the legal recognition to proceed with the process. Besides, WRS had three Director Generals and four Chairmen during the WRMP period. The institutional ownership of the policy process was shifted to as many as six ministries during the tenure of the project. These rapid changes and turnover of authority negatively impacted the institutional memory of the policy process. Although the Water Resources Management Project (WRMP) attempted institutionalization of IWRM though holistic water resources management, it failed to include a risk assessment factor and understand the changing political climate in a multi-party system of governance.

Politics of Policy Development in a Multi-party System

The WRMP project's lifespan was from 2001–06. Changes in governments during this time seriously retarded its progress. While some actors struggled to maintain the momentum of the process, others refused to support it. Different political parties that gained power during this period had different approaches to WRMP; the project stalled when those who opposed the reform process in opposition gained power. Even when the prime minister attempted to drive the policy process forward, policy development continued to be piecemeal with intermittent political interruptions.

Since the approval of the policy and institutional arrangement in 2000, the policy (revision) process had to undergo significant changes. In early 2001, the policy process was significantly affected by the creation of the Ministry of Irrigation and Water Resources Management (MIWRM). The new minister claimed "ownership" for the policy reform process and WRMP and applied pressure on the then Minister of Finance and Planning for transfer of the project and the sector reform process. The transfer process took nearly eight months and the incumbent minister instructed WRS staff not to continue with the policy development process during the handover period; this effectively delayed the policy process by eight months. The government recommenced the policy process in late 2001 but had to abandon it following renewed public agitation. A fresh policy draft was prepared, accommodating public concerns. However, in the process of revisions, the policy lost some of its key concepts of institutionalization of IWRM.

In 2002, the opposition party was elected to power and did not proceed with the reform process, having opposed the same while in the opposition. However, in 2003, an attempt was made to restart the reforms process mainly due to donor pressure to meet project covenants. This too was opposed by the public and civil society, and the reform process was abandoned. Senior bureaucrats were concerned about repeated civil society opposition. Some bureaucrats decided to present a different stance, depending on whom they were speaking to.

With increased donor pressure to achieve project milestones, the prime minister himself decided to drive the reform process, in his capacity as the Minister of Policy Development and Implementation. He was concerned that Sri Lanka should not make a bad impression on the ADB, being the second largest and most responsive donor in water development. However, it was difficult to maintain the momentum with line ministries taking a defensive attitude. In order to drive the reform process, the prime minister appointed a "task force"[21] to proceed with the reform process. The task force recommended drastic reforms: proposing separation of irrigation and water management into two ministries, on the basis that water providers and water users should not be managed under the same ministry, downsizing of the Irrigation Department and the Mahaweli Authority (saying they have completed their function as water developers and should now be management oriented), privatization of the WRB and transfer of some of the policy advisory functions to the newly created NWRA. Keeping in line with decentralization, the task force proposed establishment of a "National Water Fund" and a "Local Water Fund" to raise funds for local and national level water resources management functions. While some of these proposals had merit in making water resource management more efficient and accountable to water users, the process followed was weak and failed to build support for change. The task force limited its deliberations to a selected few members in the committee. There were no consultations with water users, stakeholder ministries or key sector institutions.[22]

The large number of actors responsible for the policy development process at different times led to more than 15 drafts of the policy and nine drafts of the Act being produced. This led to confusion and inconsistencies in contents between the policy and law. To rectify the situation, the prime minister advised his task force to re-formulate the policy in accordance with the "draft act."[23] Despite all this confusion and complexity, the prime minister was determined to take the draft act to Parliament, but had to abandon the effort due to the unexpected dissolution of Parliament by the President, requiring fresh elections.

By mid-2004, a coalition left-wing government was in power, with the support of Marxist partners. Although the need for a scientific water resources management was acknowledged by the Marxist partners, they were not in favor of continuing with the earlier reform process. The new government was faced with a dilemma, as the ADB loan had been agreed with the government of Sri Lanka as the borrower and could not be easily cancelled. The Marxist partner on the other hand did not support ADB intervention in water sector development. Subsequently, the government had to agree to a new water policy development process initiated by the Ministry of Agriculture, Lands, Livestock and Irrigation (MALLI) in which the minister was from the left-wing Marxist party. Thus, the parallel alternative water policy development process used only local inputs and no extra costs were incurred. The new "water policy" did not contain any of the key demand management clauses from the earlier ADB supported policy and was oriented towards supply augmentation.

The MALLI proceeded with the parallel policy development process using local inputs. A new draft policy was produced, promoting traditional approaches and practices. The policy was written in Sinhalese to make it more accessible and transparent to a majority of water users. The complexity of the policy development process was compounded when the new Ministry of Mahewali, River Basin and Rajarata Development (MMRBRD) was established in April 2004. The MMRBRD inherited the former policy proposals supported by the ADB. Thus the two policy processes, moving in parallel, created ill-feeling between the two ministries. Policy development was abandoned again three months prior to the presidential polls in November 2005. An attempt to renew the process in February 2006 had to be abandoned due to further adverse media publicity.

Water policy development was not consistently politically owned because of the many changes in command, shifting of responsibility between ministries (some newly created to accommodate the demands of coalition partners) and other changes. Although most of these political developments were unavoidable, better planning could have helped bureaucrats[24] play a stronger role, making the process less vulnerable to ad hoc changes. These developments clearly indicate that the policy development process was ill-conceived and lacked a clear strategy.

Politics and Institutional Sequencing

Sri Lanka has around 50 Acts and nearly 40 institutions dealing with the subject of water. The new policy proposals recommended the establishment of three new institutions, WRC, NWRA and the Water Resources Tribunal (WRT). Although this institutional arrangement was approved by the Cabinet in 2000, it was never implemented. While the WRMP continued to function, there were efforts to institutionalize this arrangement, in spite of all the other existing institutional changes to facilitate water resource management. However, some of the newly created ministries had a different approach to water resource management.

The new institutional arrangement proposed by the policy was to bridge the institutional gap identified by the earlier CWRM projects. Thus, it was believed that the new institutional arrangement would streamline water resources management, facilitate coordination and regulate water resources to improve efficiency. However, most of the functions identified for the new institutions were already present in the Water Resources Board Act of 1964,[25] though they had not been implemented. Thus, there was confusion and disagreement among water professionals regarding a new institutional arrangement.

The institutional "home" for the water policy was originally with the President[26] of the country. The policy process moved from the Ministry of Irrigation and Water Resource Management (MI&WRM) in 2001 to the Ministry of Irrigation and Water Management (MI&WM) in 2002 and then back to the MI&WRM in 2003. During this time the country's institutional arrangement for water resource management too underwent drastic changes. The country was divided into six Regional Water Resources Councils (RWRC or Water Parliaments), advising the Regional Water Resources Agencies (RWRA), who were expected to implement provisions given under the Water Resources Act and by the River Basin Organizations (RBOs). Further, the WRT, proposed under the original policy, was to be transformed to a Water Mediation Board (WMB) to arbitrate on water disputes.

In 2003, the Minister in charge of Water Management was of the view that the new institutional arrangement, especially NWRA, should create an organization large enough for him to be the Minister. As a result the original small regulatory authority, the 51-member NWRA, was to be expanded to include at least 2,000 members. Initial investigations were made for transferring assets to the newly created institution, but these could not be implemented due to subsequent political developments. However, during the time of transformation, these changes provoked the engineering fraternity in the irrigation department to protest against the creation of the new institutional arrangement, undermining the skills and expertise they had gained over a century.[27] The institutional "home" for the policy changed again in mid-2004 to MMRBRD. Although the ministry was keen to proceed with the policy process, it had to change course due to the parallel policy process initiated by the MALLI. As both policies were to deal with the same subject, the government had to resolve the problem, which threatened its political survival.

The policy developed by MALLI emphasized traditional approaches and practices, totally rejecting private sector participation in water sector development. It reinforced the public ownership of water and the state's role as the "trustee." The intention of the new policy was clearly against the earlier policy process and content and it focused more on satisfying short-term public concerns, than long-term planning. However, while the policy process was under way, the government withdrew some of the key functions of MALLI by a gazette notification and vested it with the MMRBRD,[28] so that the new draft policy of MALLI could not be submitted to the Cabinet. This created confusion and mistrust within the government ministers, leading some to allege that this, too, was a subtle attempt to facilitate water privatization.

The media, backed by a number of NGOs, highlighted the deliberate attempts within the government to subvert the new policy development efforts. Some key Sinhala language news-papers criticized the attempt as selling the country's life blood to multinationals. However, the government ignored the criticism and proceeded to obtain approval for the former policy, supported by the ADB.

In the midst of this confusion, the policy draft prepared by the MMRBRD was approved by the Cabinet of Ministers in December 2004. This led to severe criticism both by the public and ministers alike.[29] In an attempt to reconcile disputes within the Cabinet, the government appointed another committee comprising the four key ministries (MALLI, MMRBRD, Ministry of Urban Development and Water Supply and Ministry of Environment) involved in the water sector, with a mandate to review the draft proposals. However, MALLI did not send a representative to the committee, stating that they were already in the process of formulating a water resource policy.

In October 2006, the Ministry of Agriculture, Irrigation and Mahaweli Development (MAI&MD) decided to publish the draft policy for public comments. Then it appointed a committee with the four key water sector ministries to review the latest draft policy proposals. However, the committee could not meet as the ministry was changed again in February 2007 to the Ministry of Irrigation and Water Management (MI&WM). Currently, water policy formulation rests with MI&WM.

Policy Content and Controversies

Ownership

The content of the policy also contributed to the opposition it faced. Certain elements of the policy were criticized by civil society groups as being attempts to commodify water. The essence of the policy was to establish property rights to water, so that water could move from low to high productivity uses; this would also give the user full rights to ownership, while previously water was a common good according to Sri Lankan common law. Commodification may be said to require the satisfaction of three conditions: scarcity, utility and ownership (Gunatilake and Gopalakrishnan 2002*). The first two conditions were already in place in Sri Lanka and the new policy would satisfy the third. The policy, approved in March 2000, stated that "all water including surface and groundwater will be owned by the state and managed by the government in partnership with users on behalf of all Sri Lankans." This was severely contested as being contrary to common law principles: the government could transfer ownership to anybody, making water a market commodity.[30]

Equity and Efficiency

Although the policy's central principles were said to be "equity" and "efficiency" in allocation, equity concerns were masked in the body of the policy by an overemphasis on efficiency, based on demand management. This is reflected in two elements of the policy. While it advocated formal protection of water rights of bulk users through entitlements, the question of how small farmers' rights were to be safeguarded was addressed only vaguely. The proposal to shift production away from paddy said nothing about ensuring the livelihoods of the many small farmers who depend on paddy cultivation. This failure to articulate equity concerns and the lack of adequate safety nets for the poor and disadvantaged were criticized by the public as the policy was seen as attempting to provide for the "haves" at the expense of the "have nots." In general, sector reforms were not felt to be favorable to the poor and marginalized communities. The government was seen as favoring the private sector over the public sector, reflected in the privatization of the Water Resources Board and the downsizing of certain state institutions in an attempt to increase efficiency, reinforcing fears that profit would be prioritized over public good.

Entitlements

The new water policy introduced "entitlements" as a means of conferring rights to water. This terminology was based on allocations in the Murray Darling River Basin in Australia. However it was alien to Sri Lanka, and the scheme was poorly explained. For example, the policy stated that entitlements would be given only to "bulk water users,"[31] while the rights of small water users would be protected in other ways, but did not give a clear definition of either category or explain how this could be done. There was widespread speculation that bulk water users were being favored. This element of the policy was also seen by some as opening up a potential means for ownership of water resources to be transferred to multinational corporations (MNCs) in the world water market. This debate had a history extending as far back as the establishment of IWMI in 1986, which had been portrayed as a forerunner to water privatization by CSOs protesting against the water reforms. There is no real evidence to suggest that IWMI played a direct role in Sri Lanka's water policy process, but its technical and logistical support (such as the housing of the WRS) was viewed with suspicion.

The rationale for introducing the entitlement system was that it would improve the effectiveness and efficiency of water use.

However, this showed a lack of understanding of the local context. The entitlement system could be effective if the following conditions were in place:

(a) existence of a vibrant water market (demand);
(b) the necessary infrastructure to deliver the right quantity of water (water metering) to users;
(c) well developed information system for decision-making about trading of entitlements by users;
(d) an implementable regulatory system.

However, none of these conditions is currently being met in Sri Lanka. Meeting these conditions would require improvements to infrastructure to enable volumetric water allocation, the development of effective means to monitor entitlement implementation and the establishment of a new administration to manage water allocations. These would considerably increase transaction costs and risk generating corruption in granting entitlements, counteracting any possible improvements in effectiveness and efficiency. In addition, the presence of a large number of small water users makes it difficult for an effective entitlement system to work. As mentioned earlier, Sri Lanka already has a body of law dealing with water, but is ineffective in enforcement. Introducing new mechanisms and laws which are highly dependent on the same inadequate state machinery would achieve little.

Entitlements were intended to improve the equity of water allocation. Bulk water users would be issued entitlements according to demand, guaranteeing a certain quota of water. Small water users would not be issued entitlements, but instead would have their rights to use of water protected. Assuming users did not abuse the system, in theory more water would be available for everybody to share under the new approach. In practice, however, water use is highly political.[32] According to the draft water resources act, responsibility for issuing entitlements was vested with the proposed NWRA. There were some clear limitations to the process which was proposed: there was to be no credible water resources assessment to survey usage of water; the initial method for awarding entitlements was not described;[33] and it was feared that the River Basin Committees (RBCs)[34] – who were given the authority to recommend those eligible for entitlements—might be subject to political influence. Given these problems, using an entitlements system would not guarantee equity. In the midst of mounting opposition to the entitlements system, the government simply changed the terminology from "entitlements" to "permits" and subsequently to "appropriate measures." These changes pacified the public to some extent, but did not really alter the substance of the proposal and contributed to confusion and further distrust.

Permit Fee

The original draft act envisaged entitlement holders paying a "fee" for use of water. This was contentious as it was assumed to mean a volumetric charge (though this was not specified), which was seen by many as the commodification of water. The policy recommended that only bulk water users would pay, but bulk water users were not clearly defined either in the policy or the draft act and the vague definitions given suggested that Farmers' Organizations (FOs) would be considered bulk water users, leading to concerns that the small farmers accessing water through FOs would therefore also have to pay for water. The government suggested a number of different ideas to address these concerns for example, changing the term to "permit fee" (an administration fee, rather than a volumetric charge, with state water providers being exempt). However, none of this satisfied the public, as farmers were still left facing additional costs. The policy committed the government to paying the fee on behalf of farmers until their economic conditions improved, but with the increased cost of paddy production, it was still felt that the future additional costs for water would be an added burden. This was a clear case of poor attention to local context and livelihoods in the design and promotion of the new policy.

None of the earlier mentioned changes were spontaneous or accidental. They represented the thinking of the donors and were very much in line with donor policies for water resources management. These policies consider water as a commodity and advocate that conservation of water can only be achieved by fixingthe price of water. Although these issues were discussed with the bureaucracy in the WRC, none of them appears to have understood the subtle implications of these measures for water users, particularly to small farmers.

Related Reforms (Water Services Reforms Bill)

The entitlements scheme met with further resistance following the introduction of the Water Services Reforms Bill, which was submitted to parliament without public consultation. Although the bill was not directly linked to the Water Resources Act, it did conform to the overall policy environment on water sector reforms. The bill attempted to hand over the pipe-borne water supply in urban and rural areas to commercial water providers, licensed by the Public Utilities Commission, and to phase out the water supply operations of the National Water Supply and Drainage Board (NWSDB) and local authorities. The bill gave considerable power to commercial providers and aimed to introduce a licensing system for all providers, including existing state water providers. The right to a licence for existing providers, including public agencies, would be held only for five years, after which they would have to compete on commercial terms with others, including MNCs. The bill empowered commercial water providers to negotiate with state agencies and local authorities to access water directly from the source, and these negotiations did not have to answer to open and transparent public scrutiny, despite the fact that one of the negotiating partners was a public body (Rajepakse 2003). These proposals were seen as another move towards the commodification and possible privatization of water.

Confusion in Institutional Arrangements

There are a number of reasons why the three-tier institutional arrangement proposed to take over all WRM functions did not succeed. The NWRA was to be a small organization with few technical and support staff,[35] and the proposal to establish it as the apex body did not anticipate how traditional departments such as the Irrigation Department and the Water Board would react to this apparent reduction in their authority. At the same time, efforts to restructure the Irrigation Department into an "irrigation authority" to develop and manage irrigation, drainage and flood protection schemes led to confusion, as the NWRA was also expected to perform similar functions. The WRC suffered from poor political commitment and its role eventually diminished.

The WRS, established in 1996 and due to be converted into the NWRA on approval by the parliament, was supposed to act only in support of the WRC, however, eventually it was forced to take ownership of policy development, despite lacking the authority to function effectively in this role. Key water sector stakeholders opposed proposals to grant the WRS greater authority, limiting its ability to drive through reforms. In general, the policy proposals did not offer enough clarity on the roles of different institutions in the sector or adequately describe how the staff of existing agencies would be integrated into the new bodies.[36] The engineering fraternity was worried about the proposed reforms and opposed any form of restructuring in the name of water management, and some even stooped to sinister methods of opposition.[37] The failure of the proposed institutional arrangement, and the lack of clear authority granted to central institutions, was a key factor in the eventual collapse of the water policy process.

Civil Society and Media Objections

Civil society and media campaigns against the reform process were undoubtedly one of the reasons for its failure, in combination with the political difficulties described earlier. Their resistance had its roots in earlier moves towards land reform, supported by certain donors, which were seen as attempts to take away the land rights of poor farmers. Efforts to grant full land title ownership to the poor were ostensibly to enable poor farmers to use land as collateral for bank loans, but some assessments by NGOs and CSOs concluded that the real aim was to open up land markets in which land would be transferred to the highest bidder. This was seen as commodification of land, and attempts to establish water rights and the markets were seen as a dangerous extension of this policy. It was feared that they would ultimately dispossess, rather than benefit the poor, as the poor would be compelled to sell resources to escape the poverty trap.

In their opposition, civil society groups highlighted a number of policy reforms, driven by donors, which they claimed had undermined the security of developing countries, and argued that the new water policy was just another such move which would take away water from paddy farmers in the dry zone. Such criticisms of land and water policies argue that they are attacking the intractable social, economic and spiritual reliance of the peasantry on possession of land and their rights to cultivation (Seneviratne 2005). Civil society groups also wanted to ensure that any new policy would effectively protect water resources in the future. The water conservation measures imposed on bulk water users in the new policy were hailed as a step in the right direction, but the policy was silent on watershed management and control of water pollution, raising public suspicion that it was not a genuine effort to protect water resources. Due to these factors, coupled with fears of water privatization arising from events seen around the globe in the 1990s/2000s, some civil society groups came to believe and alleged that the policy was paving the way for the commodification of water.[38]

Conclusion

Water policy development in Sri Lanka is a classic case of natural resource management under a multi-party system of governance, where water has been used as a political tool. This was possible because of the poor design and formulation of the policy process. The content of the policy was not openly debated and there was no clear strategy of public consultation. This led to undermining the efforts made to institutionalize a comprehensive national policy, enabling legislation and an institutional framework to implement the policy.

A water policy of a country bestowed with strong hydraulic traditions should have articulated the values and traditional wisdom of water use; instead the policy highlighted demand management as this was the approach adopted by many donor agencies at the time for water resource management. Unable to reach a pro-poor solution to balance equity and efficiency, the policy prioritized efficiency over equity, in a country where over 60 percent of the population consists of are small farmers, depending on high water consuming paddy cultivation. This move led to severe criticism from civil society and media.

The policy advocated a strategy which had been successful in developed countries, but did not suit the local context. External policy advocates were ignorant of the political climate of the country, and failed to recognize the political nature of water and the multi-party system of governance, which would come to undermine the policy development process. The policy does not appear to have understood the value attached to paddy cultivation in Sri Lanka. Moreover, the policy process began without a clear mandate from the people, and without basic information regarding the threat of limited water resources. Thus, it was difficult for the policy advocates to convince the public of an impending water crisis, hence the need of an allocation system and a comprehensive water resource management policy.

Clearly, the policy process lacked a well articulated strategy to carry the policy formulation through. The public consultation process was not organized properly, and adequate risk assessments were not built into the process. As a result, consultations were ad hoc and piecemeal. The policy content, including the new institutional arrangement, was not clearly explained to the public. This led to confusion among the public and professionals alike, as they felt threatened by the new super structure. Policy ownership was a problem from the inception. Due to an inability to understand the application of policy content and its impact on water resources, policy ownership changed many times between ministries. This led to loss of institutional knowledge and inconsistency in continuing the process. Introduction of alien terminology like "entitlements" and "tradable rights" confused the public; they were not adequately explained, so no consensus was reached. Important documents including the approved "National Water Resources Policy and Institutional Arrangement" of 2000, were not adequately circulated and not published in Sinhala, the local language. This made the public suspicious that it was an attempt to commodify the water resources of the country. Sometimes, the government introduced policy contents through brochures, but shied away from explaining controversial elements of the policy for fear of opposition. Hence, consultation around the policy was weak, and lacked transparency and adequate dialogue.

As a result, civil society groups, who were in any case inclined to oppose the policy, saw this as an attempt to liberalize and privatize water, and believed that the livelihoods of poor farmers would be threatened. Unfortunately, the media was not adequately used in the entire process of policy formulation which is a pity because in Sri Lanka, especially, people actively use print and electronic media for information and news. As a result, both media and opposition political parties used the process to their advantage and criticized the policy, rather than focusing on the core goals of the policy and moving towards IWRM and sustainable water use.

Politicians and policy makers seriously underestimated the complexity and difficulty of the process, which had been undertaken to establish a comprehensive water policy. They did not devote adequate efforts to consult the public and communicate effectively. Besides not having a clear strategy, there was serious lack of overall commitment, which was the main cause for failure of the policy process.

Notes

1. The CWRMP consists of two technical assistance projects, Institu-tional Assessment of Comprehensive Water Resource Management (IACWRM) and Institutional Strengthening for Comprehensive Water Resource Management (ISCWRM). The strategic framework and action plan developed under IACWRM identified the need for a comprehensive policy, legislation, system of data and information management, an overarching institutional arrangement and comprehensive water resource planning in selected watersheds. The ISCWRM developed the policy, drafted the water resources act and introduced the permanent institutional arrangement for WRM.
2. It consists of two parts. Part A-capacity building to assist establishment of the National Water Resources Authority and strengthen other water-related institutions. Part B-construction of urgently needed infrastructure for water measuring, monitoring, planning and control of water.

3. IMPSA secretariat was housed in the IIMI, Sri Lanka field operations office. Some of the IIMI staff actively participated in the IMPSA deliberations.
4. The key Dublin Principles states that (a) Fresh water is a finite resource, essential for life sustenance, development and environment. (b) Water development and management should be with the participation of users, planners and policy makers at all levels. (c). Women play a central part in the provision, management and safeguarding of water and (d). Water has an economic value in all its competing uses and should be regarded as an "economic good."
5. The executing agency for part A was the Water Resources Secretariat until such time as the NWRA is established, and the executing agency for part B was the National Water Supply and Drainage Board (NWS&DB).
6. Although there is no evidence, some speculate that the reason for linking the two projects was to pull through the reform process at the expense of the much needed water management infrastructure.
7. Although Part C was a welcome addition to the project, it was intro-duced too late to the project (2004). A more comprehensive public consultation program designed by WRS in 2003 had to be aborted due disagreements between the government and the Bank.
8. The Water Resources Council and Secretariat (1999) states that 9095 percent of developed water is used for irrigation. The Ministry of Environment and Natural Resources (2002) states that more than 85 percent of the developed water is used for irrigation, 6 percent for domestic use and 5 percent for industrial purposes.
9. In the 1980s and 1990s, the construction of more than 10,000 large diameter agricultural wells in the North Central and North Western provinces have caused severe depletion of groundwater. The challenge to the Kaleni conservation barrage is due to different interpretations and understanding of the issue by the two key water management agencies. One supports the construction of the conservation barrage to prevent salt water intrusion, while the other supports construction of a reservoir to release water on demand during low flow periods.
10. Sri Lanka's population is expected to increase to 23 million by 2025 and the demand for safe drinking water is expected to be 4.6 mcm. Of this, nearly 3.3 mcm will be required by the urban population.
11. Hydro power operators control water in the upper reaches through their structures, while the WSDB allocates water at Ambatale, in the lower reaches, for domestic and industrial use. In between the irrigation department can control water for agriculture (Nanayakkara 2007).
12. The "Thuruwila drinking water supply" case in Anuradhapura. Thuruwila is a self-contained tank primarily used for paddy cultivation. An ADB-funded water supply project for Anuradhapura new town development

envisaged converting this tank into a storage tank using Mahewali waters during the wet season and feeding the new town water supply during the dry season. The capacity of the tank was to be increased, so that 21,000 m^3 per day could be drawn from it till 2020. Subsequently, withdrawals could increase up to 36,500m^3. While this was for the purpose of feeding the water supply scheme for the new town, no mention was made to the water needs of the traditional farmers who cultivated with water from the Thuruwila tank. Sector-wise, this is a conflict between irrigation and domestic water supply.

13. One of the main drawbacks of the policy formulation process was the inability or reluctance by any party to accept "policy ownership" to proceed with the process of formulation. It is inevitable that the ministry in charge of water management or water resources management should take the responsibility for "ownership" of the policy formulation, like any other water-related activity by the ministry. However, in the case of the water resources policy formulation, none of the ministries vested with the water management function took full responsibility for policy ownership.
14. Usually legal acts are not subjected to public consultation but due to the sensitivity and controversy of the proposals, it was decided to seek public opinion.
15. It is a common way of informing the public on important national policies. Normally, the information/awareness process is carried out prior to finalising the proposals. In this case, attempts were made only after the policy proposals were approved by the Cabinet.
16. Part A of the project, "Institutional Development" consists of the Establishment of the NWRA, Database Improvement and Strengthening Water Resources Management.
17. This was a major drawback of the project. Having seen the policy being approved by the Cabinet without any reservations, the Bank would have expected to use construction of the Kaleni Conservation Barrage as an incentive to approve the legislation in Parliament.
18. Although WRMP project was formally suspended in June 2006, the Ministry of Agriculture, Irrigation and Mahewali Development continued to show some interest in the water policy development process.
19. Reference to the State as the "owner" of water in sub-item 6.3.1, despite the reference to water as a "public asset" in the main item 6.2 of the draft policy of October 9, 2006.
20. Taskforce to establish the National Water Resources Authority and River Basin Committee. Although appointment of the taskforce to expedite the policy development process would have helped the establishment of the NWRA, the taskforce had no authority and its recommendations

were not binding. Some bureaucrats and technocrats in the government were critical of some taskforce members due to their previous involvement in water sector reforms.
21. One of the cardinal mistakes in the process was limiting consultations to just a few meeting with the "converts" and calling this "stakeholder consultations."
22. Ideally, legislation should follow policy. It is the duty of the Cabinet to advise the legal draftsman to prepare legislation to implement policy content. However, in this case, due to confusion and inconsistency, an attempt was made to reverse the process, but this did not materialize due to further political changes.
23. Although consultations were the mode adopted for exchanging ideas, the policy formulation process took place largely outside the bureaucratic framework. The WRS which was entrusted with the task of policy formulation, was a project based organization with added privileges and benefits. The special status given to WRS at times antagonized the bureaucracy.
24. Including: formulation of policy and planning for the water sector, conservation, utilization, prevention of pollution, integrated planning and acting in advisory capacity to the minister in charge.
25. In hindsight, although this may not have been the best approach, it appears that this was the most progressive period in the policy development process.
26. Marginalization of the irrigation department began in late 1960s, through the creation of new institutions and gradual transfer of authority elsewhere. There is evidence of a conflict between technocrats and bureaucrats in decision making.
27. Key functions of water policy formulation, formulation of river basin plans, flood and drought management, environment protection in reverie areas and management and establishment of WRS/interim NWRA, were moved from MALLI to MMRBRD.
28. Left-wing coalition partners in the Cabinet refused to approve of the policy, emphasizing that approval was granted only to appoint a committee to review the draft proposals. The Cabinet of ministers was divided on the issue and some ministers did not want to disturb the left-wing Marxist alliance, which could have jeopardizedthe existence of the government.
29. Government made several changes to the terminology with respect to ownership of water in successive policy drafts. "State ownership" became "public ownership"; then it became belongs to "all living beings," followed by a "public asset". All these changes were made to circumvent public opposition.

30. The definition of bulk water use was simply "use of water from a water source to a group of users or for a large individual user." This was interpreted as paving the way for MNCs, and large companies or farmers' companies to have rights to access at the expense of small water users. Farmers' companies were also alleged to pass the financial burden to small farmers.
31. It allows those with authority (state and non-state) to use water as a tool to vest power. The primary cause for the onset of current "Eelam War IV" (there were three uprisings between the Government and the Liberation Tigers of Tamil Elam [LTTE] since 1983) was the rebels' refusal to allow water to flow into a government-controlled area. The water policy development process partially contributed to changing the course of at least one elected government in early 2000.
32. Entitlements could be issued according to land ownership, amount of water use or by auction, none of which would protect the rights of the poor.
33. The draft Water Resources Act of 2003 proposes 60 percent of RBC (River Basin Committee) stakeholders to be non-state water users. This led to further suspicion among civil society on providing access to water to large water users.
34. It was envisaged that NWRA would consist of 51 staff, including technical and support staff. It was designed to be small as its functions were essentially coordination and management.
35. A subsequent Cabinet memo submitted by the then prime minister, reinforced the importance given to NWRA and proposed a series of reforms divesting some of the functions of key irrigation agencies to line departments and framer organizations.
36. *Waskavi* stanzas, used to curse people, were circulated through e-mail to those at WRS. The Director General of WRS died an untimely death and WRS was discontinued soon after.
37. There is a growing concern that water should not be treated as a commodity. However, it is not clear how it should be treated in the production process. While drinking and sanitation is a human right, as an input in the production process, it is a commodity. It is not very clear whether the Sri Lankan public understood these subtle differences when they protested against the policy proposals.
38. The author wishes to acknowledge the support and assistance of the Water Policy Programme of the ODI/UK, and in particular, Dr Alan Nicol, for his comments and suggestions.
39. Independent Researcher, Former Research Fellow, Agrarian Research and Training Institution and former Director, Policy and Planning, Water Resources Secretariat, Colombo, Sri Lanka.

References

Asian Development Bank (ADB). 2007. "SRI: Water Resources Management Project," Project Completion Report. Philippines: ADB.

——. 2000. "Proposed Loan to the Democratic Socialist Republic of Sri Lanka for the Water Resources Management Project". Report and Recommendation of the President to the Board of Directors. Philippines: ADB.

Bandaragoda, J. D. 2005. "Relevance of River Basin Management as a Strategy for Sri Lanka Water Sector," Consultation of River Basin Management. Sri Lanka: Lanka Jalani.

Dharmasena, G. T. 2005. "Holistic Approach to Water Resources Management", *The Island*, December 6.

Gunatilake, H. M. and C. Gopalakrishnan. 2002. "Proposed Water Policy for Sri Lanka: The Policy versus the Policy Process," *Water Resources Development*, 18(4): 545–62.

Imbulana, K. A. U. S., N. T. S. Wijesekera and B. R. Neupane (eds). 2006. *Sri Lanka National Water Development Report*. World Water Assessment Programme, UNESCO.

Rajapakse, R. 2003. "The Truth about the Water Reforms Bill," *Sunday Island*, November 30.

Saleth, M. R. and A. Dinar. 2004. *The Institutional Economics of Water: A Cross-country Analysis of Institutions and Performances*. Cheltenham, UK: Edward Elgar.

Seneviratne, G. 2005. "National Water Resources Policy", *The Island*, December 24.

Wickramage, M. 2002. "Meeting Basic Needs: Domestic Water Supply," in K. A. U. S. Imbulana, Peter Droogers and Ian W. Makin (eds), World Water Assessment Programme, Sri Lanka Case Study: Ruhuna Basins. Proceedings of a Workshop at Koggala Beach Hotel, Sri Lanka, 7–9 April.

8

Water Rights in Civil Society and Governance

The Sri Lankan Experience

Ruana Rajepakse

Overview

The legal status of water is an essential precursor to understanding water rights. How a country's legal system views its natural resources, including water, has profound implications for the rights of its citizens to such resources. The four main possibilities are:

(a) Natural resources are viewed as belonging to the state, in which case the government of the day exercises control over their use;
(b) Natural resources belong to whoever finds them;
(c) Natural resources are considered to belong to the people collectively;
(d) Natural resources cannot be owned by anyone.

In the previous two cases, it is necessary to determine who is to manage the resource, or whether it is to be left to everyone to use indiscriminately as per their wish. In this chapter, we will look at the law as it prevails in Sri Lanka, where some interesting court decisions have supplemented a fairly limited legal regime on the subject.

The Sri Lankan constitution, while including a chapter on "Fundamental Rights," does not include the socio-economic rights found in the International Covenant on Social, Economic and Cultural Rights (ICSECR), except for a few that are not related to natural resources. Nor does Sri Lanka have anything in the nature of an all-encompassing Water Act or Water Resources Act. Nevertheless, since water rights have always been important in ancient and modern societies, there is a body of common law and customary law that has been recognized and sometimes subsequently codified.

Due to an accident of history, the "common law" in Sri Lanka refers to the Roman-Dutch law. Under this system, all things are initially divided into "*res intra commercium*," that is, things for which private commercial transactions are possible, and "*res extra commercium*." The latter is further sub-divided into "*res publicae*" or things which belong to the people as a whole, and "*res commune*" which are things used by all, but belong to no one in particular.

Under these classifications, perennial rivers and harbors are said to be *res publicae* or belonging to the people as a whole, while the sea, sea shore and running water are said to be *res communes*, which can be used by all but owned by no one.[1] Thus, the water in such water bodies cannot be owned by individuals or by the state.

Management of Public and Common Resources

While neither *res publicae* nor *res communes* can be owned by any individual, various ordinances from British colonial times, as well as post-Independence Acts of Parliament have conferred rights of management and control over such things to the state. The Crown Lands Ordinance (subsequently renamed the State Lands Ordinance) conferred to the state the right of management and control of the foreshore and the water in public lakes and streams, the latter being subject to the rights of occupiers of the banks and the rights of permit holders under the ordinance.

In the case of a public stream (a river or small stream, whether perennial or intermittent, following a natural channel), the State Lands Ordinance expressly declares that the *bed* of such a public stream is the property of the state. This clearly implies that the water flowing through such a public stream is not capable of ownership, although the state is given the right to "the use and flow of the water."

It also follows that such water resources cannot be privately owned either, although riparian dwellers may have certain user rights over such resources.

Nevertheless, ambiguity has arisen over groundwater. A landowner may argue that, just as things growing on his land belong to the land and therefore to the landowner, so too does the water that is found beneath the land. However, this is a false argument as the ground water table is not demarcated to coincide with overland boundaries, and a person digging a well could end up drawing out the water under his neighbor's land and thereby depleting the ground water table for everyone in the area. This issue is yet to be clearly addressed by Sri Lanka's laws, as more fully discussed hereinafter.

Historical Background

Historically Sri Lanka was a country famed for its water management and sophisticated irrigation schemes. A medieval king once reportedly commanded that "let not even a small quantity of water obtained by rain go to the sea, without benefiting man."[2]

More importantly, in the present context, water rights in ancient Sri Lanka were well established by customs and tradition, and were based on a combination of land rights and equity. The tank or reservoir was at the center of the scheme. Cultivable land was divided into several irrigable tracts and each farmer had his own share of land within such tracts.

During times of water scarcity, the "bethma" system was applied, whereby cultivation was restricted to a smaller area that could be managed with the limited water available. Wherever this system was put into operation, existing land rights were suspended and all were required to assist in the cultivation of the smaller extent of land, from which the produce was distributed among all the farmers in proportion to their respective land holdings.

It is also noteworthy that in olden times traditional rights were perceived along with traditional duties. Under the *rajakariya* system, peasants contributed free labor for the construction of infrastructure (tanks, canals, etc.) for the irrigation system from which they all would benefit.

The British colonial administration that ruled the country from the early 19th century up to 1948 abolished the *rajakariya* system and replaced it with the *velvidana* (irrigation headman) system, under which such an official was appointed by farmers to oversee the distribution of water within the farming area and settle disputes among farmers. He was not paid a salary, but instead was allowed to collect a share of the produce from each farmer in proportion to the farmer's holding. This provided the *velvidana* with a strong incentive to ensure proper water distribution so that there would be a good crop. The colonial government also promulgated the Irrigation Ordinance, which sought to preserve the agrarian customs of the people.

Post-colonial Laws and Practices

Under the Paddy Lands Act of 1958 passed by the Government of independent Ceylon (as it was then called), the *velvidana* system was replaced by a system of "cultivation committees" (subsequently renamed "Agrarian Services Committees" in 1979). This change tends to be viewed by modern commentators as a negative development. It is alleged that farmers who were socially and politically powerful took control of the cultivation committees and arranged more favorable water distribution for their own lands. The resulting inequities led to disillusionment and lack of cooperation on the part of other farmers, while the original aim of efficient water distribution was largely forgotten. Therefore, it could be said that the opportunity for participatory management of irrigation systems was lost due to these developments.[3]

Nevertheless, there was one interesting experiment in recent times, centered around a massive multi-purpose development project using the waters of river Mahaweli, Sri Lanka's longest river. This waterway had historically been a major source of irrigation for what was then Sri Lanka's rice bowl in the north-central part of the country. Medieval wars and neglect had put this system into disrepair and disuse, and it was mainly after Independence in 1948 that attempts were made to revive the ancient irrigation network that had served the nation well in a bygone era. The most ambitious of these schemes was the accelerated Mahaweli Diversion Scheme, which was undertaken in four major phases during the late 1970s and the 1980s, and included the creation of large reservoirs for irrigation and hydropower generation. Large areas of land were also inundated in the process, and many people from hitherto fertile lands in what is known as the "wet zone" were displaced and allotted land in the "dry zone" areas that were to be irrigated by the new scheme.

In order to carry out this massive project, a special authority was created by Act of Parliament in 1979, namely, the Mahaweli Authority of Sri Lanka (MASL), which was given sweeping powers to regulate the activities under the project, including the welfare of the new settler communities. The settlements were divided into blocks, and an important experiment that was carried out as part of this scheme was a system of bulk water allocation for each block, organized as follows:

(*a*) The farmer was required to prepare a cultivation plan for his own land;
(*b*) The distribution channel farmers' organization then prepared an integrated plan for their channel system;
(*c*) The Block Level Farmer Federation thereafter prepared an integrated plan calculating the total water requirement of the Block;
(*d*) The Project Management Committee would allocate the bulk water quota.

This system had the merits of being equitable and participatory. Unfortunately, it operated only within the area gazetted under the Mahaweli Act. Only the Mahaweli Scheme and one other irrigation scheme in the south of the country were ever brought under the operation of that Act. Today the Mahaweli Act with its sweeping powers may be considered too heavy-handed, and its further use in development schemes would in any event downgrade the role of pre-existing state agencies such as the Irrigation Department and the Department of Agriculture. Bulk water allocation schemes had no statutory force in areas that were not governed by the Mahaweli

Authority, and are not recognized under the Paddy Lands Act or other agrarian legislation.

Allocation of Water for Urban Areas

It will be noted that most of the foregoing measures related to intra-sectoral water allocation, that is, allocation within an almost totally agrarian society where the only uses of water were for agricultural purposes and the domestic needs of the farming communities. The only *inter*-sectoral water allocation of any consequence was for the purpose of the hydropower schemes under the Mahaweli project, and those allocations were handled by the same authority that handled the distribution of water for agrarian purposes, namely the MASL.

With the growth of modern towns, urban water supply had become a necessity and elected municipal and urban councils were established in British colonial times. These initially enjoyed wide powers to administer public utility services in their respective areas, similar to those enjoyed by such councils in the United Kingdom. These expressly included the power to establish and maintain water supply services.

For electoral purposes, these councils were divided into wards (relatively small areas) and the people of each ward elected a member, or occasionally, in multi-ethnic constituencies, two members, to represent them in the local council. Hence there was a close link between the people and their representatives, and had this system continued, the people would have had a substantial say in how their public services, including water distribution, were administered.

However there was also a Water Supply and Drainage sub-department under the Department of Public Works, which later became a Department and, by Act of Parliament in 1974, was converted into a fully state-owned corporate body called the National Water Supply and Drainage Board (NWSDB). This Board is empowered, inter alia, "to develop, provide, operate and control an efficient, coordinated water supply and to distribute water for public, domestic or industrial purposes" in any area where it is given authority to do so by a notice published in the government gazette, by the minister in charge of the subject. The National Water Supply and Drainage Board Act empowers the Board to take over the water supply or sewerage undertaking carried on by any local council, either on the request of that council, or compulsorily if the minister, after inquiry, determines that it should do so.

The NWSDB presently carries out the operation and maintenance of more than 287 water supply schemes throughout the island, having taken over a number of schemes operated by local authorities. In 1982 it introduced water metering and billing. The board is not necessarily less efficient than the local authorities would have been (in fact the board claims to be a lot better). Nevertheless, by its own admission the 287 Water Supply Schemes operated by the Board supply only 28.6 percent of the total population with piped water supply, while another 10 percent of the population is served with hand pump tube-wells. Thus, there is still scope for local initiatives. Meanwhile, accountability, transparency and public participation in the decision-making process have been lost. Today the board is by law answerable only to the minister in charge of the subject, although as a matter of operational efficiency, it provides telephone and email services to entertain public complaints. The attitude of the board towards information disclosure can be gauged by the fact that they recently refused to release water quality data to a public interest organization that was doing a survey on access to environmental information.[4]

In the meantime, the representative capacity of local government bodies has also changed with a re-designation of many local authority areas and the introduction of a proportional representation voting system, with each party fielding a list of candidates for the entire local authority area instead of designating a candidate for each ward (constituency). This has severed the direct link that formerly existed between a council member and the voters of his ward. Recent surveys have shown that the new system is very unpopular, and hence any move to give back old powers to the new-fangled local authorities may not evoke much public interest, or may even be opposed. An all-party Select Committee of the Parliament has recommended that the present proportional representation system that operates at all levels (the Parliament, provincial councils and local authorities) should be replaced by a mixed system under which two-thirds of the seats of each such body would be directly elected and the remaining one-third be allocated on a proportional basis amongst the losing candidates. At the end of 2008, parliamentary bills were laid before Parliament to introduce this proposal at the local authority level.[5]

Subject to the power of Parliament to alter the voting system as outlined earlier (which of course alters the power of the people), local authorities are actually in a strong position vis-à-vis the national government after the Thirteenth Amendment was passed in 1987, which declares that the existing powers of local authorities may not be diminished. As will be seen later in this chapter, this constitutional provision has proved significant in the context of moves to reform the water sector.

Resolution of Inter-sectoral Water Disputes

On paper Sri Lanka is not considered a water-scarce country. However, there are frequent water scarcities in parts of the country due to significant variations in rainfall, while much of the rainwater that falls is allowed to flow into the sea unutilized, despite the injunction of the medieval king quoted at the beginning of the chapter. Increase in population and greater industrialization are also increasing the demand on water resources, while extraction of groundwater for commercial purposes remains virtually uncontrolled. In addition, existing water resources are threatened by pollution and excessive sand and gem mining from riverbeds.

While agriculture has traditionally been the largest water user in the country, competing demands from other sources are increasing and thus a mechanism for equitable inter-sectoral water allocation is a pressing need.

However, this need is not easy to fulfil, firstly because there is a debate over what is equitable. Some believe that traditional uses, particularly agriculture, should be given precedence, while others advocate the most cost-efficient use of water. There is also an increasing conflict between the needs of rural and urban communities competing for water from the same supply sources.

The determination of what is equitable in these circumstances could take place in one of the following ways:

(a) A national policy to determine the order of priorities.
(b) Allocation to be determined by market forces.
(c) A special adjudicating body or bodies may be established to settle water disputes.
(d) It may be left to concerned citizens to seek redress from the regular law courts or other dispute resolution bodies, such as mediation boards.

In Sri Lanka thus far, only the fourth method referred to in the preceding points is available. Even under this head, examples are few, and the best known are the "Eppawela" and "Thuruwila" cases, so called because of their respective locations. Both cases were filed under the fundamental rights jurisdiction that allows aggrieved citizens to directly petition the Supreme Court.

The first of these cases concerned the proposed commercial exploitation of a phosphate deposit at Eppawela in Sri Lanka's agriculturally rich North Central Province, which contains some of the country's most important modern and traditional irrigation systems. Hence, any large scale pollution of this area would have had a serious impact on water resources and food security. Phosphate from this deposit was already being exploited at the rate of about 40,000 metric tons (MT) per year and local scientists were of the view that the country could sustain an increase of up to about 1,50,000 MT per year, sustainability being important, as phosphate is a non-renewable resource. However, the Government of Sri Lanka was about to enter into an agreement with a multinational conglomerate to carry out a high intensity mining operation to mine 26.1 million MT within 30 years, which would have exhausted known reserves in the lifetime of a single generation and caused massive environmental pollution in the process. The petitioners were persons of the area whose lands and livelihoods, including the water sources to irrigate their lands, were threatened by this project, for which no environmental clearance under Sri Lankan law had been obtained.

The Supreme Court held, firstly, that the fundamental rights of the petitioners to choose their occupation and their place of residence had been infringed by the proposed project. While under Sri Lankan law, land may be lawfully acquired by the state for a public purpose, the petitioners successfully argued that the project under consideration was so damaging that it could not reasonably be considered as a "public purpose." In addition the court held that the petitioners' right to equality before the law and equal protection of the law and been infringed because no environmental impact assessment as required by Sri Lanka's National Environmental Act had been carried out prior to approving the project.[6]

The Thuruwila case, decided in 2002, was the first recorded case in the Supreme Court relating specifically to water rights and inter-sectoral water transfers.[7] The petitioners were rice farmers who had traditionally cultivated their lands with water from the Thuruwila Tank, a self-contained rain-fed reservoir, once again located in Sri Lanka's agriculturally rich north central Province. An Asian Development Bank (ADB) sponsored water supply scheme for the nearby Anuradhapura New Town proposed to convert this self-contained tank into a storage basin to accumulate water from river Mahaweli during the rainy season and feed the New Town during the dry season. The capacity of the tank was to be increased so that 21,000 cubic meters of water per day could be drawn from it up to the year 2020 and the draw-out increased to 36,500 cubic meters thereafter. No mention was made as to how the water needs of the Thuruwila farmers were to be met.

As in the Eppawela case referred to earlier, the farmers went to court on the basis of an imminent infringement of their fundamental rights to equality and equal protection of the law, and to engage in any lawful occupation of their choice. The court, conscious of the water needs of both the farmers and the town folk, encouraged the parties to formulate a scheme that would cater to the interests of both parties. The terms of settlement thus entered illustrated the degree of practical detail that is often required in judgments on economic rights. The main terms of the settlement were as follows:

(a) The Mahaweli Authority would as far as possible, ensure a daily input of 27,000 cubic meter of water into the tank;
(b) The maximum daily draw-out from the tank would not exceed 21,000 cubic meter;

(c) In any event the daily draw-out would not exceed the input;

(d) The daily input and draw-out would be monitored by instruments and a record would be maintained, which would be accessible to the petitioners and other members of the public;

(e) A sum of INR 2 million was to be made available to compensate the petitioners for any losses suffered as a result of the project;

(f) In the event of damage being caused to the tank bund in the execution of the project, the petitioners would be entitled to pursue additional claims for compensation from the relevant authorities.

These are not the only occasions when the Supreme Court, acting at the behest of aggrieved citizens, has intervened in water-related disputes. There is an ongoing case which was filed by leaders of three community-based organizations in 2006. It concerns the issue of virtually unregulated sand mining from a river bed in the northwestern part of the country. It had not only destabilized the river banks, but had also caused the level of the water to drop to such an extent that the water supply scheme for the main town of the area was threatened. In fact, sand mining requires a licence, but the allegation of the petitioners was that there was a widespread flouting of the law in which the miscreants were backed by local politicians and the police were therefore turning a blind eye. The court, having granted an injunction against any further sand mining from the river in question, has assumed the role of monitor, whereby the police and officers of the mining regulatory authority are required to report to the court every two–three months. The petitioners are allowed to mention any shortcomings and the respondents are required to state the remedial action they have taken. As a result, the petitioners have now been able to raise funds and embark on a project to rehabilitate the river banks. Amongst the project's workforce are several poor people of the area who previously had to make a living doing the illegal mining for the racketeers.

This case, popularly known as the Deduru Oya case after the river in question, highlights two interesting developments:

(a) The willingness of local communities, with capable community leadership, to work together towards protection of water resources.
(b) The willingness of the Supreme Court to play a monitoring role where other agencies of the state have failed.[8]

Lack of a National Water Policy

It is now necessary to consider why the other options referred to above for the resolution of inter-sectoral water disputes have not materialized, namely:

(a) A national policy to determine the order of priorities;
(b) Allowing market forces to determine water allocation;
(c) Establishment of a special adjudicating body or bodies to settle water disputes.

The rationale for having a National Water Policy is perhaps best expressed in the preamble to India's National Water Policy as follows:

> Water as a resource is one and indivisible; rainfall, river waters, surface ponds and lakes are all part of one system.[9]

Sri Lanka has a public institution, namely the Water Resources Board, whose responsibility specifically includes advising the minister on "the formulation of national policies relating to the control and use of the water resources of the country." It is a multi-disciplinary body which may include persons from outside government service, and there is also an Inter-Departmental Advisory Committee which strengthens the inter-sectoral aspect of the Board. However, perhaps owing to another provision in the Water Resources Board Act specifically dealing with "the proper control and economic use of groundwater," this Board has now effectively been confined to dealing with groundwater resources, and that too, in a non-regulatory capacity, as the Board is an advisory body only.

In Sri Lanka, the only Cabinet approved water policy was the one approved in 2000, which had been prepared by a special Secretariat created for the purpose and funded by the ADB. This document, along with a parallel draft "Water Resources Authority Act" ran into considerable public controversy. It made the proposition that all water resources are "owned" by the state. This was contrary to the common law concepts about water as set out at the beginning of this article, and also appeared to depart from the public ownership and/or public trust principle that had hitherto characterized the recent decisions of the Supreme Court, also referred to earlier. Devoid of the concept of public ownership or public trust, such a provision would have given the right to the government of the day, representing the state, to transfer the ownership of water resources to whomsoever it chose. The fact that the Secretariat which drew up this policy was ADB-funded helped to fuel suspicions that transfer of ownership or "privatization of water" was the real agenda.

An emphasis on cost recovery principles and a provision to allocate water on the basis of transferable water entitlements, as well as references to the need to "balance consumptive water uses such as irrigation, domestic and industrial, and non-consumptive uses such as hydropower, fisheries, recreation and aesthetics" led to a perception that the proposed measures were designed to favor private sector economic activities over traditional livelihoods, particularly agriculture. Furthermore, under the section of the policy on "demand management," Sri Lanka's rice farmers who had hitherto been seen as important providers of food security to the nation, were to be encouraged to turn to less water-intensive crops. Demand management was described in the policy as "increasing the economic efficiency of water use by allowing water to be moved to higher valued uses over time."

Critics of the policy appeared to see it as a paradigm shift away from the hydraulic civilization of the past to a new culture that saw water as a commercial commodity to be utilized in the most "economically efficient" way possible. This Cabinet approved policy was never officially rescinded, but neither was it implemented, mainly due to the pressure of public opinion. A redrafting process began, and during 2002–05, a number of draft policies and draft Acts were in circulation, though not formally notified to the public. The most controversial aspects of the 2000 policy, including the language, had been significantly modified, although the proposed institutional arrangements remained the same.

Thereafter on October 9, 2006, a full page notice was published in the press by the Ministry of Agriculture, Irrigation and Mahaweli Development containing a draft "National Water Resources Management Policy" and inviting public comment. This was the first time that a draft National Water Policy was published in the media and comments invited from the general public, although this practice has been followed in the case of many other national policies including, the National Wetland Policy that was prepared by the Ministry of Environment and adopted by the Cabinet in 2004.

The key concepts and aims of this 2006 draft policy may be summarized as follows:

(a) sustainable use of water resources;
(b) goal of a balanced environment conducive to social and economic development;
(c) water recognized as a basic need of all living beings;
(d) water for domestic and sanitary needs to be given priority;
(e) groundwater extraction to be monitored and regulated;
(f) river basin to be hydrological unit for planning and management of water resources;
(g) gender concerns recognized;
(h) "Water rights" to be recognized and transferable;
(i) a participatory approach to be followed,
(j) demand management;
(k) "Economic efficiency and accountability of water use;"
(l) watershed management and catchment area protection;
(m) date and information management;
(n) institutional arrangements that include:

   1. National Water Resources Management Authority (NWRA)

2. National Water Resources Management Council (NWRC)
3. Water Resources Management Tribunal (WRT)
4. River Basin Committees (RBC)

Most of these key concepts were in harmony with existing legal principles and likely to be welcomed by all those who understand the need for proper water management and conservation. In particular, the proposed River Basin Committees were a welcome move to re-introduce participatory decision-making, although care would be needed to ensure that their powers did not run foul of the quasi-federal arrangements between the government and the provincial councils introduced into Sri Lanka's Constitution in 1987.[10]

However the following proposals were likely to attract controversy:

(a) The reference to water rights and in particular *transferable* water rights: As indicated at the beginning of this chapter, water rights have existed under various laws and customs in Sri Lanka from time immemorial. Amendment of these rights as and when necessary is permitted by due process of law, just as some water rights may be lawfully curtailed, for instance, by conservation and anti-pollution measures. However, the choice of words in the draft policy was strongly suggestive of a move to turn water into a marketable commodity. This is entirely different from merely charging a reasonable price for water services as is presently done, and would be likely to run foul of modern human rights concepts, as well as traditional common law principles.

(b) The reference to the state as the "owner" of water in one sub-clause of the draft policy, despite the reference to water as a "public asset" in another place, appeared to be a throwback to the policy document of 2000.

(c) The complex institutional arrangements envisaged in order to give effect to the policy would be an additional public expenditure which a cash-strapped country like Sri Lanka could ill afford. Hence there would be a risk that such institutions would be under-resourced and under-staffed, which would bring the proposed scheme into disrepute. Such arrangements were also likely to resurrect fears about a "hidden agenda" – an allegation never far from the surface when this topic is discussed among the Sri Lankan public. In any event a national water policy should apply to all institutions that handle water, and the creation of a special body to implement the policy may discourage other water-related agencies from taking common ownership of the policy.

With regard to the proposed Water Resource Management Tribunal, there is certainly a need for a quick and inexpensive forum for resolving water disputes. However, as illustrated by the Thuruwila case, such disputes are likely to occur at the local level and involve persons of relatively low income and education levels. There is a need for easily accessible mechanisms at the local level, perhaps guided by a head office staff to ensure that uniform standards and criteria are applied.

However, it must also be said that there is in Sri Lanka a vocal lobby that appears to be opposed to any attempt to regulate the water sector. It is significant that nothing further has been publicly heard of the 2006 draft policy, although more than two years have elapsed since the closing date for public comments.

Water Issues and Human Rights

On November 26, 2002, at the close of its 29th Session, the United Nations(UN) Committee on Economic, Social and Cultural Rights issued General Comment No. 15 of 2002, declaring that:

> Water is a limited natural resource and a public good fundamental for life and health. The human right to water is indispensable for leading a life in human dignity. It is a prerequisite for the realization of other human rights.

The publication of this General Comment followed a practice commenced in 1988 whereby the UN Committee on Economic, Social and Cultural Rights prepares "general comments" on the rights contained in the International Covenant on Economic, Social and Cultural Rights (ICESCR) with a view to assisting state parties to fulfil their reporting obligations and to provide greater interpretative clarity as to the intent, meaning and content of the ICESCR. According to the Office of the UN High Commissioner for Human Rights, the Committee viewed the adoption of such General Comments as a means of promoting the implementation of the Covenant.[11]

Acknowledging the fact that water is not expressly mentioned in the ICESCR, the Committee based its stand on the provisions of Articles 11 and 12 of the ICESCR. Noting that Article 11(1) of ICESCR specifies a number of rights emanating from, and indispensable for the realization of the right to an adequate standard of living "including food, clothing and housing," the Committee held that the use of the word "including" indicated that this catalogue of rights was not intended to be exhaustive. It went on to state that the right to water clearly fell within the category of guarantees essential for securing an adequate standard of living. Article 12 declares the right of everyone to the enjoyment of the highest attainable standard of physical and mental health.

The Committee, which comprises international legal experts who function in their individual capacity and not as representatives of member states, accordingly set out a series of practical guidelines for the implementation of this concept which may be summarized as follows:

(a) water should be physically accessible and economically affordable;
(b) under no circumstances should an individual be deprived of the minimum level of water essential for life;
(c) the water needs of subsistence farmers and indigenous peoples should be recognized;
(d) national standards should be developed for water quality, using World Health Organization (WHO) guidelines;
(e) while it is recognized that countries face resource constraints, there should not be inappropriate allocation of resources, e.g. on expensive schemes that benefit a few, as against those that benefit the larger part of the population;
(f) the state should protect the water rights of the public against arbitrary interference from third parties in the form of denial of access, pollution or inequitable extraction of water resources;

(g) traditional systems of water allocation should be respected;
(h) the right to water should be given legal recognition and there should be an effective regulatory system;
(i) there should be access to information and public participation in the decision-making process;
(j) the state should adopt a national strategy and plan of action for ensuring that water is available and affordable to everyone, and establish competent institutions to carry out this strategy and action plan.

Sri Lanka is a signatory to the Covenant, even though many of the socio-economic rights set out therein are not expressly included in the country's Constitution. However, as already indicated in the cases reviewed in this chapter, the Supreme Court has used its powers under the chapters on Fundamental Rights, and in particular Article 12 which guarantees equality before the law and equal protection of the law, to secure economic and environmental justice for the people in the cases that have been brought before it.

In a Supreme Court case in 2003, where three citizen petitioners challenged certain clauses of an Intellectual Property Bill designed to comply with the requirements of the Trade Related Intellectual Property Rights (TRIPS) Agreement in Sri Lanka, the court declared that developed and developing countries "cannot be treated as equals" and held that the Bill would be inconsistent with the equality provisions of the Constitution, unless certain mitigatory measures such as compulsory licensing for drug production to meet public health emergencies were incorporated into the Bill.[12]

Similar principles would no doubt be applied to water issues and it is unlikely that the courts would countenance a policy of allowing market forces to determine water supply. In fact, shortly later that year, when the government sought to introduce a Water Services Reform Bill, the petitioners in three cases filed against the Bill, argued that the Bill violated Article 12(1) by being heavily weighed in favor of commercial water service providers, as against the consumer.[13]

The passage of the Bill was in fact stopped by the court on a procedural ground, namely, because the Bill sought to phase out the power of local authorities to run water services, and was thereby inconsistent with the Constitution, which states that the powers of local authorities cannot be taken away. Hence, the continuing importance of the local authority in relation to water issues, as mentioned earlier in this chapter.

This case also marked the first occasion on which petitioners had an opportunity to cite the UN General Comment No. 15 on Water referred to earlier, and although the court halted the Bill on the procedural ground as referred to in the preceding paragraph, it recorded the petitioners' submissions on the human rights issue in its judgment:

> [The petitioners submit] "That, in terms of the General Comment No. 15 (2002) of the Economic and Social Council of the United Nations, the right to water comes within Articles 11 and 12 of the International Covenant on Economic, Social and Cultural Rights. The General Comment specifically states that the right to water is to be considered a human right, and since Sri Lanka has acceded to the International Covenant, the provisions of our law be interpreted in the light of obligations under the Covenant and in terms of the General Comment made thereon by the Council. On that basis it is contended that the right to water would come within the purview of fundamental rights as referred to in Article 3 of the Constitution and form part of the sovereignty of the People. That the Bill which seeks to remove the function of providing pipe-borne water supply to the people by state agencies including Local Authorities and vest it in private sector licencees would amount to an alienation of the Sovereignty of the People inconsistent with Article 3 of the Constitution."

The Degree of Private Sector Participation in Water Services

Sri Lanka has from time to time received foreign assistance for the setting up of water infrastructure, to which there has been no audible objection. However the matter of maintenance, supply and billing in the urban sector is handled entirely by the National Water Supply and Drainage Board referred to earlier in this chapter, while the Irrigation Department or other relevant public authorities handle the maintenance of irrigation schemes and distribution of water for irrigation.

It is interesting to note that the UN General Comment on water does not rule out private sector participation in the supply of water, and in fact appears to assume the existence of this possibility. However the General Comment makes clear that no matter who distributes the water, the state has an obligation to ensure that it is equitably distributed, even to those who cannot afford to pay for it.

The following provisions of the General Comment are relevant in this context:

(a) The human right to water entitles everyone to sufficient, safe, acceptable, physically accessible and affordable water for personal and domestic uses (Paragraph 2);
(b) Water and water facilities and services, must be affordable for all. The direct and indirect costs and charges associated with securing water must be affordable, and must not compromise or threaten the realization of other Covenant rights (Paragraph 12(c)(ii));
(c) Any payment for water services has to be based on the principle of equity, ensuring that these services, *whether privately or publicly provided*, are affordable for all, including socially disadvantaged groups. Equity demands that poorer households should not be disproportionately burdened with water expenses as compared to richer households (Paragraph 27, emphasis mine).

Theoretically, as long as water is affordable, it should not matter who supplies and distributes it. However, it is arguable whether a commercial water supplier can meet its own commercial needs and still supply water at the same price that a public agency could do, for the following reasons:

(a) While a public water service, if it is to run efficiently, needs to recover its operational costs and make a reasonable profit to enable the periodic upgrading of plant and services;
(b) A private corporation doing the same thing needs both the above mentioned elements plus an added profit margin to pay dividends to its shareholders.

Even if water for the poor is subsidized, there is a hidden cost to the public because it is the state that has to meet the cost of the subsidy. This operates in a particularly insidious manner under a regime of privatized water services, as the state has to pay the private water service company to compensate it for the loss it suffers in providing subsidized water. Such a provision is contained in Public Utilities Commission of Sri Lanka Act of 2002 which was introduced by a pro-privatization government to regulate the water and electricity industries, but was never implemented as the said government was defeated in a general election in 2004.

In any event, the reputation of most of the leading water companies worldwide is so bad that any attempt to introduce them into Sri Lanka would be likely to meet with considerable civil society resistance. This appears to be tacitly acknowledged by the present government (2009) which has initiated action to bring electricity distribution under the Public Utilities Commission, but has yet made no such proposal regarding water services.

It should also be noted that private sector involvement in water services is not the same as "privatization" of water itself, which is not possible under the common law of Sri Lanka (and indeed most countries).

Groundwater Extraction

Groundwater is said to account for more than 85 percent of safe drinking water in the rural areas of most Asian countries, including Sri Lanka. Due to its drought reliability, good natural quality and continuous availability, groundwater has become immensely important for human water supply. In Sri Lanka, it is known to be used for homesteads as well as hotels and by water bottling companies. However it is still "insufficiently understood, widely undervalued, irrationally exploited and inadequately protected."[14]

Groundwater extraction is therefore one area where local "privatization" of water is de facto being permitted. As mentioned earlier in this chapter, the Water Resources Board, though mandated to advise the minister on ground water issues, has no regulatory powers. The environmental impact assessment procedure that operates under the National Environmental Act applies only if the extraction exceeds half a million cubic meters per day.[15] Not only is this a very high threshold, but it is not practically possible to ascertain whether it has been exceeded, unless there is a system to licence and monitor all extraction of water from underground sources. Unfortunately, attempts to do so have met with emotive objections from misguided sections of civil society who perceive any such move as an attempt to tax the poor people who draw water from a well. As a result, there is no legal control or even monitoring of groundwater extraction and the water bottling companies are reported to be flourishing.

Protection of the Natural Environment

If water rights are to be meaningful, it is essential that the natural environment be protected to the full extent necessary to ensure the continuance of a supply of fresh potable water, as well as the preservation of rainfall catchment areas. Conservation measures and anti-pollution measures need to be rigorously enforced.

The Constitution makes clear the shared obligations of the state and the citizens in this regard. Under the Directive Principles of State Policy, the state "shall protect, preserve and improve the environment for the benefit of the community." Likewise, it is one of the "fundamental duties" of the citizens to "protect nature and conserve its riches."[16]

Sri Lanka has a plethora of laws dealing with land, water and environmental issues and a multitude of agencies to enforce them. This may be considered as one of the weaknesses of the system, as often each agency is found to be doing its "own thing." However the principal law is undoubtedly the National Environmental Act which was passed in 1980 and greatly strengthened by an amending Act in 1988. The principal features of the Act as amended are:

(a) Establishment of the Central Environmental Authority (CEA) as the principal regulatory and project approving agency.
(b) The requirement that any state agency or private developer has to obtain environmental clearance before embarking

on any "prescribed project" (a list of such projects being gazetted and amended from time to time).

(c) The introduction of Initial Environmental Examination (IEE) and Environmental Impact Assessment (EIA) procedures for approval of projects.

(d) The introduction of public participation in the approval process. However by an amendment passed in 2000 such participation was limited to projects that are deemed to require EIA and not those that are approved only on the basis of an IEE. There are no published guidelines to indicate which types of projects may be approved by IEE and which ones should be subjected to EIA.

(e) The requirement that no one shall discharge waste into the environment, without a licence obtained after prescribed procedure.

(f) The introduction of the "polluter pays principle."

(g) The establishment of an Environmental Council which, together with the CEA, is empowered to recommend to the minister the basic policy on the management and conservation of natural resources.

(h) A provision that in the event of inconsistency between the National Environmental Act and any other legislation, the former shall take precedence.

It has to be said that the CEA is yet to perform to its true potential, but thanks to the information disclosure provisions in its enabling Act, it has emerged as one of the more public-friendly institutions. However its performance in the water sector is particularly weak. While it has taken some actions to stop pollution of water bodies by industries and garbage dumping, it is yet to enforce the "polluter pays principle" under which the polluter may be required to rehabilitate the polluted water body at his own cost, or the CEA may do so and charge the cost to the polluter. Furthermore, in many instances, the pollution or other wrongdoing has been highlighted by public interest groups filing writ applications, naming the CEA and other relevant parties as respondents. In many such cases the CEA has adopted a cooperative attitude in court and the cases have been settled without going to final judgment. However this does not always ensure proper rehabilitation of the natural ecosystem that has been damaged, which can only be ensured by strict enforcement of the "polluter pays principle", for which a court judgment would be required.[17]

The CEA has powers under the National Environmental Act to assume a more proactive role with regard to policy formulation and law enforcement. Nevertheless, governmental ambivalence with regard to water issues has prevented the CEA, and the broader-based consultative body known as the National Environmental Council, from being given the responsibility of policy formulation in the water sector, except for the drawing up of a National Wetlands Policy, which was put up for public comments and received the approval of the Cabinet in 1994.

Conclusion

In conclusion one has to ask – Where does all this leave the citizen and the state?

Regarding the citizen, the answer, in the Sri Lankan context, is not entirely negative. The foregoing chapter has demonstrated that it is public spirited citizens who have been at the forefront of most of the initiatives to preserve and equitably distribute water resources, aided by a largely pro-public judiciary. The rules of *locus standi* have been liberally interpreted, and this writer is unaware of any environment-related case where a public interest litigant has been denied legal standing to sue.[18]

In the absence of specific water laws and policies, the emphasis tends to be on equity. This includes the principle of equitable distribution, as in the Thuruwila case, and conservation in the interest of inter-generational equity, which was a concept specifically cited in the Eppawela judgment and is also referred to in respect of policy formulation under the National Environmental Act. In addition, the Constitution gives space for laws that violate fundamental rights including the principle of equal treatment, to be challenged (prior to enactment) by any citizen.

Justice based purely on cases is not satisfactory for two reasons – firstly, it is reactive rather than proactive, and sometimes irreparable harm has already been done. Secondly, there will always be instances where the affected people are too poor or insufficiently aware of their rights to initiate action, and public interest groups cannot maintain a nationwide vigil.

Nevertheless, in the absence of any policy or long-term strategic intervention by the state, the courts may be the only forum available for citizens to vindicate their rights and ensure the rights of future generations in regard to water. Such judgments will also have to be respected by those misguided sections of civil society that are implacably opposed to any form of water regulation by the state on the grounds that such interventions have a hidden agenda.

Notes and References

1. This analysis is based on Voet 45.1.6 and 1.8.4 to 1.8.8, as set out by Peiris, G. L., 'The Law of Property in Sri Lanka,' Volume 1, pub. Lake House Investments Ltd.
2. *Mahavamsa* 68.8-13, translated by Mudaliyar L. De Zoysa, *Journal of the Royal Asiatic Society* (C.B.), Volume III No IX and cited by Sri Lanka Supreme Court in *Bulankulama and others v. Secretary, Ministry of Industrial Development and others*, [2000] 3 Sri Lanka Reports 243 at 254.
3. For a critique of the system see M. I. M. Mowjood and S. Pathmarajah. 2007. "Review of Historical Water Sector Coordination and Allocation Issues: Prospects and Problems for the Future", pub. Sri Lanka Water Partnership proceedings of Consultation on Effective Water Institutions to Promote Integrated Water Resources Management.
4. The legal powers of the Board are governed by the National Water Supply and Drainage Board Act No. 2 of 1974 as amended, while the details of its operations are gathered from its website: http://waterboard.lk/scripts/ASP/adout_us.asp (accessed on November 10, 2013). The refusal to release information is documented in a NGO report: *The Access Initiative (TAI), Sri Lanka Report (draft) – Access to Environmental Information, Public Participation and Access to Justice,* September 2007.
5. Interim Report of the Select Committee of Parliament on Electoral Reforms, Parliamentary Series No. 08, June 5, 2007; and Local Authorities Elections (Amendment) Bill 2008.
6. *Bulankulama and others v. Secretary, Ministry of Industrial Development and others*, [2000] 3 Sri Lanka Reports 243. The judgment in this case also made reference to the UN Declarations of Stockholm (1972) and Rio (1992) and the 1987 Report of the United Nations Commission on Environment and Development (UNCED) Report chaired by Gro Harlem Brundtland.

7. *H.B. Dissanayake and 8 others v. Gamini Jayawickrema Perera, Minister of Irrigation and Water Management and 5 others*, S.C.F.R. 329/2002 decided on September 30, 2002 (unreported, as it eventually took the form of a consent order rather than a judgment).
8. *Gunatilake and 5 others v. Hon. Maithripala Sirisena, Minister of Agriculture, Irrigation, Mahaweli Development and Environment and 9 others*, S.C.F.R. 226/06, instituted 26.06.2006. The constant monitoring by the court negated the influence of political thugs and racketeers in the area to a large extent, although sporadic acts of violence were reported to court even in 2007.
9. Clause 1.2 of National Water Policy India, issued by Government of India Ministry of Water Resources, April 2002. This Policy is an update of the National Water Policy that was adopted in 1987 and contained the same phrase.
10. The Thirteenth Amendment to the Constitution (1987) established elected Provincial Councils which were vested with significant powers over land and water within each province. Provincial boundaries do not coincide with the boundaries of the proposed river basin areas.
11. UN Economic and Social Council E/C.12/2002/11, 26 November 2002: General Comment No. 15 (2002): The right to water (articles. 11 and 12 of the International Covenant on Economic, Social and Cultural Rights). The purpose of General Comments is more fully set out in UN Fact Sheet No.16 (Rev.1), Committee on Economic, Social and Cultural Rights, issued by The Office of the High Commissioner for Human Rights, Geneva, Switzerland.
12. Supreme Court Special Determinations Nos.14-16/2003, reported in 7 S.C.S.D. 34, pub. Citizens' Trust.
13. Supreme Court Special Determinations Nos. 24 and 25/2003, reported in 8 S.C.S.D. 35, pub. Citizens' Trust.
14. Panabokke, C.R., 'Groundwater conditions in Sri Lanka – A Geomorphic Perspective' pub. National Science Foundation of Sri Lanka, 2007.
15. Ministerial Order issued and gazetted under National Environmental Act.
16. Articles 27(1) and 28(f) respectively. Although the Directive Principles of State Policy are declared to be not enforceable in any court or tribunal, no part of the Constitution can be treated as irrelevant and such principles should at least be an aid to interpretation of what the law requires. Nevertheless, Sri Lanka has not built up a body of case law on the significance of the Directive Principles as has happened in India. By contrast, the Fundamental Duties of the citizen, which also include a duty to "uphold and defend the law" are regularly cited by public interest petitioners as the basis of their legal standing to file suit.
17. Writ applications are filed against public bodies or holders of public office to quash an unlawful decision or prohibit an unlawful act or compel

the public body or person to perform a public duty. Such actions may be filed by affected persons or by persons acting in good faith in the public interest. In Sri Lanka writ applications are separate and distinct from fundamental rights cases and are utilized with equal frequency by public interest groups. Unlike fundamental rights applications where the petitioner is required to show that his or her rights have been, or are about to be, infringed, a writ application may be filed by a person whose own rights are not affected but who is seeking to uphold the law in the public interest.

9

"Flood Action Plan" and NGO Protests in Bangladesh

An Assessment

Hamidul Huq

Overview

Water is life in Bangladesh, a country that has more than 200 rivers, 16,000 sq. km. of wetlands, and 80 percent of whose area is under floodplains (Chowdhury et al 1997)[1] that support rich ecosystems, both in quantity and diversity and are a category of riparian zones or systems. The annual surface water flow through Bangladesh is impressively large, and ground water is one of the major natural resources. Bangladesh is a delta of 147,570 sq. km ., in the lower basins of the three great rivers – the Ganga, Brahmaputra and Meghna. Floods are a recurring phenomenon and take place almost every year in Bangladesh. This annual phenomenon has played a vital role in shaping the economy, society and culture of the country (ibid.). It is usually said that "moderate floods" have occurred once in two years on an average, while "severe floods" have occurred once in 6–7 years on an average (Huda 2005). However, these trends vary. For example, severe floods occurred in consecutive years in 1954 and 1955, and in 1987 and 1988. Floods in Bangladesh are caused mainly by overflows from the major rivers – Ganga, Brahmaputra and Meghna, and their tributaries and distributaries from June to September. Also, the huge inflow of water due to rainfall in the upstream catchments, as well as different parts of the country that cannot be drained out due to high stages in the outfall rivers leading to river bank overflows, breaking dykes and flooded villages, crop fields, towns and cities. An outfall river is the discharge point of other rivers, where it discharges into the sea. The river Meghna is the outfall point of all other rivers in Bangladesh through which they discharge water into the Bay of Bengal.

There have been 16 flood disasters in Bangladesh in the last 50 years, which caused a damage of BDT 70,000 crore (Bangladeshi taka), estimated as equivalent to five years' revenue expenditure of Bangladesh. The flood of 1955 caused an inundation of 39,000 of 147,570 sq. km. of the national territory. The flood of 1988 inundated 102,000 sq. km. and directly affected 40 million people in a population of 100 million. The consecutive devastating floods of 1987 and 1988 killed about 1,500 people. Crops of 4 million hectares of land were destroyed, 3,000 km of roads and 2,500 km of embankment and 1,900 water control structures were either completely destroyed or badly damaged (Siddique 1989). The floods of 1987 and 1988 caused extensive damage in all the sectors and severely disrupted the economy, reducing potential Gross Domestic Product (GDP) by about 4 percent (Flood Plan Coordination Organisation 1995).

Monowar Hossain (1994) notes that studies based on documentary evidence of floods were not available until a report prepared by Prasanta Mahalanobis for the Irrigation Department of Bengal in 1927. After the catastrophic flood in north Bengal in 1922, Prasanta Chandra Mahalanobis, a professor at Presidency College in Calcutta, conducted a study on records of floods between 1870–1922 for the Irrigation Department of Bengal. This study observed that the river system had undergone profound changes, important channels had silted up and drainage had deteriorated considerably. Mahalanobis concluded that, "Embankments may, for a time, prevent overflow from the rivers, but would tend to raise the bed of the rivers still further, and thus make the situation much worse in the long run."

Following a series of devastating floods in 1954 and 1956, the United Nations (UN) commissioned a Technical Assistance Mission headed by J. A. Krug. The Krug Report noted that the problem of controlling floods in East Pakistan (now Bangladesh after the 1971 Liberation War) was so difficult, and available engineering data so meagre, that no clear solution could be recommended (cited in Hossain 1994). While recommending the construction of flood control embankments, for instance, Krug was concerned about its adverse affects and recurring liabilities (ibid.). One can argue that politics implicitly and explicitly entered the Krug Mission Water and Flood Report. This became more explicit in 1964, when the government, drawing upon the Krug Mission recommendations, charted out a 20-year Master Plan of 58 projects, with an estimated budget of US$2 billion, for the water resources development, by hiring foreign and local consultants. The Master Plan proposed building thousands of kilometers of embankment projects and was linked to the Green Revolution programs of the then Military Government of Bangladesh, supported by the United States Agency for International Development (USAID). The Green Revolution program attempted to double and triple cropped area by irrigation and the protection of cultivable lands by flood control and drainage schemes. But, in fact, production gains on land inside the embankments tended to be short lived (ibid.). As Hossain (1994) and Chowdhury et al. (1997) noted, hundreds of years ago, people used to construct earthen embankments around cultivable land to protect it from flood water. The construction of dykes along rivers and irrigation by planned excavation of canals were practiced during Mughal rule in the 16th–18th centuries. The zamindars engaged people to build embankments along the rivers in order to protect crops from the flooding of rivers. The zamindars charged local farmers a long-term levy for an embankment by requisitioning a part of their harvest, and the farmers were forced to maintain the embankments. However, long before the embankments and polders, that is, an area of low-lying land reclaimed from inundation by building of dykes(there are about 123 polders in Bangladesh) in recent decades, rural communities built earthen walls around their rice fields to protect crops from tidal intrusion in the coastal zone and the usual flooding in flood-prone areas of Bangladesh (Hossain 1994). But the dominant development interventions in water sector, especially in flood control projects, predominantly guided by expatriates' recommendations and donors' preferences, brush aside these sustainable local practices. Thus, water and politics are closely connected in the water-centric development interventions guided by dominant actors – from the zamindar to the state, expatriates and donors and lenders like the Asian Development Bank (ADB) and World Bank.

Birth of the Flood Action Plan

The devastating floods of 1987 and 1988 had induced international communities to help Bangladesh. Among them was Danielle Mitterrand, wife of former French President, Francois Mitterrand, who while visiting Dhaka as a guest of the then Bangladesh President, Hussain Muhammad Ershad witnessed and was greatly moved by the floods which took place in 1988. The French president, in search of projects to implement his vision of greatly expanded aid to the Third World, responded by sending a team of 30 French engineers to Bangladesh, to study the flood problem with Bangladesh government engineers and to draw up solutions (Boyce 1990).

Several studies, shortly after the 1988 floods, were also conducted with the technical and financial involvement of institutions like the United Nations Development Plan (UNDP), as well as nations including Bhutan, China, India, Japan, Nepal, United Kingdom (UK) and the United States of America (US). These studies mentioned that the technical, economic and environmental information available was inadequate to make optimal choices to address the flood problem in Bangladesh. Nonetheless, these studies recommended several projects (Flood Plan Coordination Organisation 1994). In fact, the team of French engineers prepared a most dramatic, long-term proposal in 1989 recommending the construction of hundreds of kilometers of long embankments along the great rivers of the Bangladesh delta, including the Ganga and Brahmaputra, initially costing US$ 5.2–10.1 billion. Mitterrand proposed this plan at the G-7 Paris summit in 1989. General Ershad also keenly pursued a major "flood control" initiative with international funding agencies, with an enthusiasm attributable in no small measure to the prospect of lucrative construction and middleman contracts for Bangladeshi firms (Boyce 1990). Accordingly, the G-7 leaders and the Ershad government asked the World Bank to coordinate flood control plans for Bangladesh (ibid.). The World

Bank coordinated efforts to produce a program of studies and pilot projects that formed the basis for a long-term comprehensive flood control plan. It convened a meeting in Washington in 1989, attended by a Bangladesh government delegation and leading experts involved in the initial studies, and proposed an action plan for the next five years as the first step towards formulating a long-term program (Flood Plan Coordination Organisation 1994). In December 1989, the World Bank presented its Action Plan for Flood Control, named "Flood Action Plan Bangladesh" (FAP), in the "development partners meeting" held in London, attended by representatives of the US, Japanese and European governments, who committed US$150 million to the Flood Action Plan in the initial phase (Boyce 1990; Flood Plan Coordination Organisation 1994). Despite grave and well-founded reservations about the wisdom of massive embankments, the long-term strategy envisioned in the World Bank's action plan incorporated central elements of the French scheme (Boyce 1990). The mega Flood Action Plan commenced in 1990.

Flood Action Plan as the Panacea

The Flood Action Plan was proposed as the panacea for all flood problems in Bangladesh. In fact, as dictated by the G-7 leaders, it was highly donor-driven and dependent on experts. This mega flood control project initially had 29 components, of which 18 were studies and pilot projects in its first five-year phase 1990–94. The Plan objectives were to safeguard lives and livelihoods; minimize potential flood damage; improve agro-ecological conditions for higher crop production; meet the needs of fisheries, navigation, communication and public health; promote commerce and industry; and create a flood-free land for a better environment.

In 1990, the Government of Bangladesh also set up the Flood Plan Coordination Organisation (FPCO), which had no legal status, to supervise, coordinate and monitor Flood Action Plan activities, aided by a "Panel of Experts" and specialists from home and abroad. The government also set up technical committees and multidisciplinary teams, with government and donors' representatives, for technical reviews of project proposals, reports, and approval (FPCO 1995). All these arrangements were under control of the World Bank, the coordinator of Flood Action Plan, which guaranteed absolute domination and manipulation in the designing and decision making of the projects.

At the end of the first phase of the Flood Action Plan, the FPCO produced the "Bangladesh Water and Flood Management Strategy" document in March 1995, endorsed by the Cabinet and the donors. It recommended a five-year (1995–2000) program focusing on the following:

(a) preparation of a national water management plan;
(b) strengthening of water sector organizations responsible for planning, construction, operation and maintenance;
(c) implementation of high priority projects. It recommended a US$7.12 billion investment in 162 projects over five years. Overall, its objective was "to provide tools and a framework for a wide range of investigations into issues relating to water management, river training, flood mitigation and drainage."

The Protests Against the Flood Action Plan

From its inception till the end (1990–2000), the Flood Action Plan encountered criticism, opposition and conflict by the locals, politicians, government officials, professionals, academics, researchers, intellectuals, students, activists, media, and Bangladeshi and international NGOs. Given its precariousness, asymmetry, donor-driven agenda, expatriate dominancy, non-transparency, heavy physical engineering approach, and the controversial reputation of its coordinating agency, that is, the World Bank, the Flood Action Plan faced enormous hurdles throughout its tenure. The NGOs and locals opposed the Flood Action Plan's authority and its primary stakeholders–the experts, donors, consultants and contractors.

The first opposition to the Flood Action Plan was in Tangail, 100 km north of Dhaka, by the locals, especially women, fishermen and farmers. They opposed the Flood Action Plan component called Compartmentalization Pilot Project (CPP). This component was popularly known as FAP 20, because, the serial number of this component was 20. The FAP 20 was funded by the Dutch,

German and Bangladeshi governments, respectively and implemented by the Bangladesh Water Development Board (BWDB), in partnership with Bangladeshi and Dutch consulting firms. A "compartment" was described as "a component surrounded by embankments, with gated openings on the upstream side and openings without gates mainly at the downstream side, through which the inflow and outflow of flood water can be controlled. A system of channels and *khals* (canal) has the function of transporting the water inside the compartment." The goal of the CPP was to increase productivity in agriculture and fisheries and secure rural and urban development through "controlled flooding," by creating 17 sub-compartments in the project area, commanding 13,200 hectares. But it received a lot of criticism over the structures and water regulators (Kemik et al. 1998). For instance, the Flood Action Plan staff used the example of the Netherlands to convince the locals, but later there was violence when dealing with the contractors.

There was widespread opposition to the Flood Action Plan, led by NGOs through their apex body, the Association of Development Agencies in Bangladesh (ADAB), Coalition of Environmental NGOs (CEN), Forum of Environmental Journalists in Bangladesh, Bangladesh Citizens' Group and others. They demanded the Plan's accountability to the local people, its alignment with national interest, its commitment to a people-centric policy, advocacy with the government and donors and recognition of the NGOs as development partners of the government.

The Flood Action Plan was opposed on the following grounds:

(*a*) Its process not only excluded people's participation, but involved anti-people activities;
(*b*) Local people had been harrassed during its implementation;
(*c*) Its embankments were a death trap;
(*d*) Land was forcibly acquired for its implementation and compensation not paid; if paid, it was not legitimate;
(*e*) Information about the plan was not shared with the people;
(*f*) The Flood Action Plan was illegal under the existing laws of Bangladesh;

(g) It was never discussed in the Parliament of Bangladesh;
(h) Many flood control projects failed to work and in fact, intensified the flood;
(i) It focussed on flood control, without a holistic plan that considered river erosion and drought concerns;
(j) Engineering structures caused water logging problems;
(k) The Flood Action Plan portrayed water as enemy of Bangladesh, whereas Bangladeshis consider water as life and a potential resource ;
(l) It ignored and undermined local people's knowledge in water resource management.

The key demands of NGOs in the Flood Action Plan protest campaign were:

(a) Stop all Flood Action Plan activity immediately;
(b) Have an independent review and evaluation of Flood Action Plan activities;
(c) Make available an audit and complete evaluation of the finances and expenses of the Flood Action Plan till date;
(d) People's participation at all stages of water resources management;
(e) Full compensations for all the people affected by the Flood Action Plan;
(f) Integrated water resources planning and management;
(g) Ensure regional cooperation in water resources planning and management;

Flood Action Plan Protest Strategies

The Flood Action Plan protest movement opposed the irrational, unprofessional and undemocratic behavior of the Bangladesh government, donors and the World Bank. It demanded a holistic approach, including integrated water resource management and policy, people's participation, environmental impact assessment, water resources management plan, sustainable development, water and gender, river dredging and an efficient institutional arrangement for implementation. The Bangladeshi NGOs worked for greater access to policy information, particularly concerning the Flood Action Plan, and dialogue with the policy/decision makers. They built networks with national and international media, environmental activist groups, the Dutch Parliament, the European Parliament, research organizations, lobby groups, academic institutes, other NGOs for a greater and strengthened platform and international NGO donors. They made Flood Action Plan information available to the locals, as well as emphasized the value of local expertise and experience in traditionally dealing with floods, to the Bangladesh government and donors.

M. M. Rahman, Executive Vice Chairman of the Bangladesh Agricultural Research Council, stated in 1989 that the Bangladeshi public has "the incontrovertible right to know about the costs and benefits of the proposed flood control investments." However, democratic accountability has not been a prominent feature of development planning in Bangladesh. He was promptly dismissed from his government post (Boyce 1990).

The other Flood Action Plan stakeholders displayed undemocratic behavior as well. For instance, at a meeting, which I attended, organized by the Berlin-based Information, Alternatives and Opposition (IAO) Network in 1994 in Berlin, a video on the protest demonstration in Tangail in September 1993 against Flood Action Plan was presented. Dr Kinneman and Jorn Kreischer, representatives of the German KfW (Kreditanstalt fur Wiederaufbau), that is, *Reconstruction Credit* Institute, a German government-owned development bank, based in Frankfurt, expressed "clear doubts" and accused the organizers of the Tangail demonstration of having misinformed the people about the consequences of the project. If they had been "told more about the real project", they would not have been demonstrating, they insisted.

Likewise, during a Dutch Parliament debate on the Flood Action Plan on February, 10 1994, the Minister for Development Cooperation, asked why he did not suspend construction under Flood Action Plan when it was opposed by the locals, said there were interests at the stake beyond those of the local population.

It can be argued that the governments of donor countries often ignore the interests of the local people whom they assist in developing countries, in order to protect their own interests. No norms, democratic values or rationality seem to be mandatory when "imperialist" actors take decisions, affecting those in developing nations, of which the Flood Action Plan is a glaring example.

Plausible Policy Influence

The NGOs were assertive about their success in the Flood Action Plan protest movement. They had built linkages with international NGOs, environmental activists, media and lobbyists, and got access to information, policy debates and politicians regarding the Flood Action Plan policy. They influenced the European media to raise a critical mass of dissent against the Flood Action Plan, as it was funded by their tax payers' money, but against the interests of the Bangladeshis. The Bangladeshi NGOs were partly dependent on international NGOs, for instance, to secure access to the International Water Tribunal, obtain support from the US-based International River Network, mobilize the Green Group's opinion on the issue in the European Parliament, and adaptation of the June 1993 European Parliament resolution, which urged member states to reconsider their support to the Flood Action Plan. This was largely the result of lobbying by Dutch NGOs and the Berlin-based Information, Alternatives and Opposition (IAO) Network, which coordinated the international Flood Action Plan protest campaign (Majumder 1994).

The NGOs' campaign generated debates on the Flood Action Plan in politics and civil society. Many international NGOs were convinced that their people should know whether their taxpayers' money was causing damages to interests of the people of developing countries by the wrong decisions of their politicians and the experts.

Probe International, a Toronto-based group that monitors the impact of Canadian foreign aid and trade, based on their observations, described, Flood Action Plan (FAP) as "a mind-boggling scheme designed to deny the fact that 80 percent of Bangladesh is floodplain." In Probe's opinion, FAP's US$10 billion will sink the country deeper into debt and further increase its dependency on external assistance at the expenses of local initiatives. They wanted both the World Bank and Canadian International Development Agency (CIDA) to abandon the Flood Action Plan (Majumder 1994). But the Flood Action Plan protest movement was conflicted on the issue of donor support and how to proceed further. Some American NGOs like International River Network (IRN), National Audubon Society (NAS) criticized the World Bank, which was coordinating the Flood Action Plan, on the grounds that the Bank's general strategy was to impose mega-projects, so that it could lend large amounts and perpetuate indebtedness. The World Bank sponsors the world's worst environmental, social and economic disasters in the name of aid to the Third World (Barber et al. 1993). While some North American NGOs pressurized the World Bank and other donors of the Flood Action Plan to pull out, European NGOs tended to concentrate more on trying to make their own governments more accountable. Some groups wanted immediate withdrawal; others wanted support for the scheme to be made conditional on certain improvements (Majumder 1994). And although Bangladeshi NGOs were sharply critical of the Flood Action Plan, many did not regard withdrawal of support from the World Bank and other donors as the solution. They were worried that if the Bank stopped coordination and donors withdrew, the aid money would be redirected into bilateral water projects in an uncoordinated fashion, which may lead to free-for-all. (ibid.).

Among the debates on the Flood Action Plan, one high-level debate took place in the European Parliament in May 1993, with the European Parliament taking a critical position on the FAP (IAO-Network 1995). Another was a statement of concern on the Flood Action Plan by several British NGOs like the international British NGO, OXFAM, Christian Aid and German NGOs like EZE, WEED, Urgewald and Bread for the World to all donors in early 1994, that called "for a suspension of all construction activities until the Flood Action Plan met some crucial conditions like (a) the full participation of all affected people in project planning and implementation was ensured, (b) a comprehensive environmental impact assessment was completed, (c) the social, legal and economic rights of any people to be resettled were fully respected." Also, an alliance of 21 Dutch environmental and development organizations submitted a petition to the Dutch Minister for Development Cooperation, prior to a Dutch parliamentary debate on the Flood Action Plan, demanding that all construction work on the Flood Action Plan be suspended, including construction work under pilot schemes (ibid.). A meeting of Dutch Members of Parliament had with Flood Action Plan consultants, reviewers, European Flood Action Plan Platform members and NGO representatives from Bangladesh (including the author) in the Hague in 1995, prior to the international donors' conference on the Flood Action Plan in Dhaka, was expected to be considered as crucial in influencing the Dutch government to rethink the Flood Action Plan. A "Dutch Flood Action Plan Platform" arranged meetings between Bangladeshi NGO representatives (including the author) and Dutch Parliament Members. The NGO representatives also met Wilfried Telkamper, a German member of the Green Party in the European Parliament, who had visited the Flood Action Plan 20 area in Tangail and shared his concerns with the government and donors. National and international papers and journals in Germany and the Netherlands covered the people's protest with articles critical of the Flood Action Plan.

As a result, the Dutch and German governments commissioned reviews of Flood Action Plan 20, but the review report was kept confidential. Following a letter from the Dutch Minister for Development Cooperation to its members of the Parliament dated January 30, 1996, in response to the parliamentary debates on Flood Action Plan, Bangladeshi NGOs prematurely claimed success in influencing policy, until they realized that in Dutch the letter meant that the government was going ahead with the Flood Action Plan. In his parliamentary speech, the Dutch Minister for Development Cooperation said, "I know there is local opposition, but we will have to continue the project in greater interest." He added that on the basis of positive developments, the recommendations of the Overseas Inspection Department (IOV) of the Dutch government and the policy, it is within reason for the Netherlands to continue its participation in the Flood Action Plan beyond 1995 (ibid.).

Thus, one can argue that Dutch politicians and the donor government deliberately decided to continue funding the Flood Action Plan, brushing aside Bangladeshi interests, in order to further their own. Similarly, bilateral and international donors and moneylenders like the World Bank often make financial investments and exert political power to bully dissidents and protect their own interests and hegemony. A professor of Geography at the University of Manotoba and an expert on China's energy and environment expressed outrage upon seeing the feasibility study of CIDA and the World Bank funded project, "Three Gorges Dam" in China, saying, "This is not engineering and science, merely an expert prostitution, paid for by Canadian taxpayers" (Barber et al. 1993). Moreover, the foreign-funded flood projects in Bangladesh were partnered by the Bangladesh government. Thus, the Flood Action Plan is one example where colonial rule has been replaced by development cooperation policy, well after Bangladesh's independence.

Transposed Strategy

At the start of the second phase of the Flood Action Plan (1995–2000), the Bangladesh government, the World Bank, donors and experts continued to face criticism. An international panel of experts, including engineers and environmentalists, set up to oversee the Flood Action Plan, condemned the "final report" (the World Bank draft of the report entitled "The Flood Action Plan, Final Report") of first phase of Flood Action Plan, written by Bangladeshi engineers with the help of World Bank officials. The panel said that the Flood Action Plan was over-ambitious and ignored key environmental impact, including, damage to the river fisheries that provide most Bangladeshis with their protein (*New Scientist*). In mid-1994, the Flood Plan Coordination Organization submitted their first proposal for 700 projects costing over $7 billion. Following a World Bank review, the list was reduced to 162 projects, but still cost $7 billion (*The Financial Express*, Dhaka, November 30, 1994).

"We've read the report (of the Flood Plan Coordination Organization) thoroughly, and it really doesn't make the case for the plan," wrote Jim Dempster, a British engineer. He noted that many issues that emerged in the research were not addressed in the report (*New Scientist*). Meanwhile, a United Nations Development

Program (UNDP) review commission found the Flood Plan Coordination Organization's report to be full of holes, and recommended further studies to clear doubts about its envir-onmental and socio-economic impact, and taking full account of the views of the people to be directly affected by the projects (*The Financial Express* 1994).

A World Bank official, William Smith, who wrote part of the draft report of the first phase of the Flood Action Plan, noted that a large program of embankments would be "truly disastrous," but claimed that the Bank had "transformed" the plan into something manageable and environmentally friendly. This draft report suggested that the dispute was far from settled and now threatened the entire future of the Flood Action Plan (*New Scientist* 1994). Following his visit to the Flood Action Plan 20 area in Tangail, Wilfried Telkamper, German member of the Green Party in the European Parliament, said in Dhaka that the fate of the Flood Action Plan should be decided not only by policy planners and donors, but by the Bangladeshi people and the parliament (*Daily Star* 1994).

Importantly, the Flood Action Plan had never been discussed in the Bangladesh Parliament. During the discussion on the UNDP report, the World Bank mission chief in Bangladesh, Mr. Pierre Landell-Mills categorically stated that the Bank would not finance controversial projects. The UNDP report noted that, as the Flood Action Plan was a controversial project, it should not proceed further until the controversies were resolved (*Bhorer Kagoj* 1995).

Despite all the opposition and debate, the second phase of the Flood Action Plan (1995–2000) was launched by the Bangladesh government. The plan, renamed the Bangladesh Water and Flood Management Strategy, was adopted by the Cabinet on September 11, 1995 (*Green Watch* 1995), with complete donor support and some NGO support. It has been suggested that some donors found usage of the Flood Action Plan tag troublesome and disguised the same, old-style water sector interventions under new names (Huq and Rahman 1994). Many projects came to Bangladesh with other names, for instance, Southwest Area Integrated Water Resources Management Planning project – a replication of Flood Action Plan 20. Philip Williams of the International River Network, speaking at a news conference of the International Water Tribunal, Amsterdam in February 1992, says that, "Bangladesh's debt situation is extremely bad, and the World Bank has a vested interest in churning over the debt, in finding excuses to give money to Bangladesh so that it can repay its old loans. It does this by creating disastrous projects to pay off the loans for old development projects – part of the debt cycle" (Majumder 1994).

However, during the second phase of Flood Action Plan, the Flood Action Plan advocacy strategy of the Bangladeshi NGOs and their overseas partners changed significantly. The Bangladeshi NGOs' strategies were variously seen as proactive, educative and partnering, as well as co-opted and compromising.

The Northern (Europe and North America) NGO partners and Coalition of Environmental NGOs (CEN) of Bangladesh endorsed the Final Report of the Flood Action Plan (FAP) entitled, "Bangladesh Water and Flood Management Strategy," which was officially submitted to the Prime Minister of Bangladesh in September 1995, as an outcome of the Flood Action Plan protest campaign. The CEN expressed in their statement on the documents that, "the general emphasis on future Integrated National Water Planning in an environmentally sound and socially acceptable manner is welcomed. CEN has all along advocated such an integrated approach towards water sector planning rather than over-emphasis on Flood prevention." The CEN also noted that, "the member NGOs of CEN have substantial experience and expertise in developing environmentally sound natural resource planning with participation of the poor and would be able to offer that experience and inputs into the proposed National Water Planning exercise." In the same statement, the CEN also noted that, "the decision to continue with the pilot construction projects identified in the first phase of FAP, namely FAP 3, 20 and 21 without completing the National Water Plan is wrong and should not be supported" (CEN 1995). In 1996, the "Dutch NGO Flood Action Plan Platform," the Europe-based, strongest lobby partner of the Bangladeshi NGOs, wrote to them, saying that, "If the NGO movement intends to capitalize on its political clout in the Flood Action Plan debate, the Bangladeshi voice needs to be most prominent. The Dutch NGO Flood Action Plan Platform therefore feels that the role between the Bangladeshi network and itself has gradually changed. It sees that the major share of the activities (monitoring Flood Action Plan implementation) and the responsibility to react to this process (will be) the portfolio of the Bangladeshi NGOs." They clarified their own role as restricted to responding to requests from Bangladesh to act on certain developments by addressing these directly to the northern (the donor countries) decision-making authorities, and channeling Northern information to Bangladesh. They offered workshops in lobbying skills and campaign techniques to Bangladeshi NGOs." The Bangladeshi NGOs felt that their northern NGO partners were evading lending support to them; some northern NGOs were dependent on government funding. The World Bank also distanced itself from the Flood Action Plan tag. The Bangladeshi NGO protest movement soon faded, and the Flood Action Plan went unchallenged in Bangladesh.

One could say that it is business that primarily guides water sector projects in Bangladesh – business for multilateral moneylenders, bilateral donors and expatriates, as well as local consultants, construction firms and line agencies. They use power of all kinds – monetary, political, even linguistic (hired expatriates produce tons of documents using the latest development jargon that the layman cannot understand). They ensure that investments pay, even at the cost of the local peoples' livelihoods, destruction of ecosystems and increased intensity of flood disasters.

Conclusion

The protest movement against the Flood Action Plan originated from the local people's opposition to the Plan's activities, particularly the engineering infrastructure, such as Flood Action Plan 20 in Tangail, Flood Action Plan 21/22 in Gaibandha and Flood Action Plan 3.1 in Jamalpur district. The local people first came to know of the Flood Action Plan only when contractors began structural work, or foreign consultants were spotted in the area. People opposed the Flood Action Plan after their opinions were ignored. Bangladeshi NGOs involved in the Plan protest movement built strong networks at home and abroad, which gathered some momentum. But they were hampered by a lack of key inputs, including access to Flood Action Plan information, engineering expertise in water and flood control, funds for campaign activities including public debates and studies, and limited access to political power, business, academia and the government water sector. They were also dependent on donors, and eventually fizzled out.

About 700 Flood Action Plan recommended projects are being implemented without institutional reform, which was a key demand of donors and national stakeholders during the Flood Action Plan protests. The Flood Action Plan continues to expand its projects in the Bangladesh water sector through evasion of responsibility, and lack of accountability and transparency from donors, experts and decision makers. The water projects, especially the flood control projects, are not designed using local knowledge and experience of flood and water management, ecosystems, agriculture, environment and local institutions. They are designed by foreign experts, ignoring the local peoples' opinions and local knowledge base. They are dictated by dominant internal and external forces protecting their own vested interests within and outside Bangladesh, but not in the national or local people's interest.

Meanwhile, the project's name is no longer Flood Action Plan. The World Bank has pulled out from Flood Action Plan but they are continuing to lend money in several water sector projects. Many projects have become bilateral and multilateral. For example, the "Southwest Area Integrated Water Resource Management Plan" project is supported by the ADB and the Dutch government, the Water Management Improvement Project (WMIP) is supported by the Dutch government through the World Bank, and many others (Annual Report of Water Resources Ministry 2007–08). At present, the country has a mess of uncoordinated bilateral and multilateral water projects. The water sector organizations of the Bangladesh government like Water Resources Planning Organization, Bangladesh Water Development Board and the Local Government Engineering Department are implementing water sector projects individually with the bilateral and multilateral funding support.

Bangladesh attracts massive aid in the form of loans from the World Bank and the ADB. For instance, in 2007, the ADB approved a total loan amount of US$ 965.7 million out of which US$ 200 million was to help improve the management and operations of urban water supply systems. The Local Government Engineering Department (LGED) alone implemented 332 Small Scale Water Resources Development Projects during the fifth Five Year Plan (1997–2002). As of 2003, nearly 600 schemes with various combinations of flood control, drainage, irrigation and salt water intrusion mitigation objectives, owned and operated by the Bangladesh Water Development Board (BWDB) covering a total areas of about 6 million ha, almost 60 percent of the net cropped area of the country. The past investment by BWDB in water projects has been about US$ 3.4 billion up to the year 2002 (ADB 2003).

The water sector has also been identified as the most corrupt one. The Anti-Corruption Commission (ACC) of Bangladesh's investigations in the water sector have revealed that corruption in different projects run by the Ministry of Water Resources was estimated to be approximately US$ 1.5 billion during 2001–06 (Suryanarayanan 2009).

Nonetheless, the NGO protests had a long-term positive impact leading the government to formulate a comprehensive plan for water and flood management, a national water policy and a national water management plan. The various NGOs are partners in many of the projects as well.

Note

1. A floodplain is a flat or nearly flat land adjacent to a river that stretches from the banks of its channel to the base of the enclosing valley walls and experiences flooding during periods of high discharge. In other words, a floodplain is an area near a river which inundates or floods easily. Flood plains are made by a meander eroding sideways as it travels downstream. Geologically, ancient floodplains are often represented in the landscape by fluvial terraces.

References

Boyce, K. James. 1990. "Birth of a Megaproject: Political Economy of Flood Control in Bangladesh, in 'Environmental Management'," 14(4): 419–28.

Chowdhury, J. U., M. R. Rahman and M. Salehin. 1997. *Flood Control in a Floodplain Country: Experiences of Bangladesh*. Rabat: Islamic Educational, Scientific and Cultural Organization.

Comments on Bangladesh Water and Flood Management Strategy of Flood Plan Coordination Organisation by Coalition of Environmental NGOs (CEN), November 1995. Unpublished Document.

Flood Plan Coordination Organisation, Report on The Flood Action Plan, an unpublished document of the Flood Plan Coordination Organisation, Ministry of Water Resources, Government of Bangladesh, Dhaka, October 1994.

Hossain, Monowar. 1994. "Existing Embankments," in K. Haggart (ed.), *Rivers of Life*, Dhaka: BCAS.

Huda, A. T. M. Shamsul. 2005. "Integrated Water Resources Management in Bangladesh: An Assessment," in Asit K. Biswas, O. Varis and C. Tortajada (eds), *Integrated Water Resources Management in South and Southeast Asia*. New Delhi: Oxford University Press.

Majumder, Mostafa Kamal. 1994. "Activists' Responses," in K. Haggart (ed.), *Rivers of Life*. Dhaka: Bangladesh Centre for Advanced Studies (BCAS).

The Daily Star, national daily, Dhaka, March 5, 1994.

The Financial Express, national daily, Dhaka, November 30, 1995.

PART IV
Social Movements in Water Governance

10

The Water Rights Movement in South Maharashtra, India

Suhas Paranjape and *Seema Kulkarni*

Civil Society, NGOs and Social Movements

The term civil society, as it is largely used today, has undergone a shift in meaning from its original historical sense. The term is used in its clearest form by the German philosopher Georg Hegel to distinguish it from and hold it in counterpoint to the state (Hegel 1820). Hegel distinguishes it from the state and family, and includes in it all forms of association outside these two spheres, including the economic sphere. Although Karl Marx initially adopted this usage, he later criticized Hegel, and reached radically different conclusions regarding the role played by civil society. In the writings of Marx, civil society splits into the capitalist economy and bourgeois society. For Italian Marxist philosopher Antonio Gramsci and German sociologist Jurgen Habermas, it broadly denotes bourgeois hegemony, the exercise of dominance through consensual and non-coercive forms while the state represents the exercise of dominance through the use of force or actions based on the threat of force.

Shift in Meaning: The 'Third Sector'

The question of whether or not civil society includes the economic sphere – markets and businesses – has not always been clear and is primarily attributed as the reason for the shift in meaning. This is illustrated by the non-governmental organization (NGO) protests that erupted when a United Nations (UN) website provided links to its relations with both businesses and NGOs from a single civil society web page (Willets 2002). This shift in meaning may be summarized in the London School of Economics' Center for Civil Society definition: "Civil society refers to the arena of uncoerced," collective action around shared interests, purposes and values. In theory, its institutional forms are distinct from those of the state, family and market, though in practice, the boundaries between state, civil society, family and market are often complex, blurred and negotiated. Civil society commonly embraces a diversity of spaces, actors and institutional forms, varying in their degree of formality, autonomy and power. Civil societies are often populated by organizations such as registered charities, development non-governmental organizations, community groups, women's organizations, faith-based organizations, professional associations, trade unions, self-help groups, social movements, business associations, coalitions and advocacy groups (Centre for Civil Society 2004).

In short, though there are considerable difficulties in separating civil society from the state on the one hand, and the market and the economy on the other, what is clear is that it denotes the entire social space that lies between the state and the economy – the "third sector."

NGOs

The NGOs and social movements are located in the social space known as the "third sector." The historian Akira Iriye defines an NGO as, "a voluntary, non-state, non-profit, non-religious and non-military association. The voluntary nature of these organizations – that is, their openness to all those who wish to join – distinguishes them from others that are restricted to certain categories of people, whether nationality, gender, religion, class, region, or any other division" (Iriye 2002). Professor Willets describes an NGO as, "an independent voluntary association of people acting together on a continuous basis, for some common purpose, other than achieving government office, making money or illegal activities" (Willets 2002).

Definitions of NGOs have to deal with two demarcations: firstly, to demarcate the objectives of NGOs from the objectives of other types of organizations, and secondly, to demarcate them from others in the manner of their association. The latter is not a problem since it is the voluntary association that distinguishes NGOs from businesses and state organizations. But NGOs can have a diversity of objectives and a European Commission discussion paper summarizes the difficulties of defining NGOs, while offering a good description:

> The NGO sector has often been described as extremely diverse, heterogeneous and populated by organizations with hugely varied goals, structure and motivations. It is therefore not an easy task to find a common definition of the term 'non-governmental organization'. It cannot be based on a legal definition, given the wide variations in laws relating to NGO activities, according to which an NGO may have, for instance, the legal status of a charity, non-profit association or a foundation. The term NGO can nevertheless be used as shorthand to refer to a range of organizations that normally share the following characteristics: NGOs are not created to generate personal profit. Although they may have paid employees and engage in revenue-generating activities they do not distribute profits or surpluses to members or management; NGOs are voluntary. This means that they are formed voluntarily and that there is usually an element of voluntary participation in the organization; NGOs are distinguished from informal or ad hoc groups by having some degree of formal or institutional existence. Usually, NGOs have formal statutes or other governing document setting out their mission, objectives and scope. They are accountable to their members and donors; NGOs are independent, in particular of government and other public authorities and of political parties or commercial organizations; NGOs are not self-serving in aims and related values. Their aim is to act in the public arena at large, on concerns and issues related to the well being of people, specific groups of people or society as a whole. They are not pursuing the commercial or professional interests of their members (Prodi and Kinnock 2001).

Much of the description applies to what earlier used to be known as voluntary organizations (VOs) and the shift in usage to NGOs may be related to two possible factors. Firstly, the term NGO can be traced to the United Nations (UN), which, because of the nature of its mandate, had to make a clear and explicit distinction between those organizations, government and inter-government that fell fully within its mandate and those that did not. The increasing globalization of issues, actual and perceived, is likely to have tilted usage in favor of the UN term. Secondly, the earlier VOs had before them a model – irrespective of how far the situation corresponded to the model – in which voluntary activity was largely self-financed or voluntary, and aid and funding were supplements. The shift in usage is also likely to be related to a model in which NGO activity is largely dependent on institutional funding, whether state or private.

Social Movements

However difficult it may be to pin down precisely what an NGO is, as formal organizations, NGOs have clearly delineated organizational boundaries and clearly defined formal objectives. Characterizing social movements are a much more complex matter, even without entering into the whole issue of the "new" social movements. It is often acknowledged that the popular usage of the term is quite loose. As discussed in Allwood and Wadia 2000:

> The expression "social movement" has been and continues to be used quite loosely. Social movements are said to be collective in nature; organized outside established political and corporate institutions precisely because those involved lack regular access to such institutions; and contentious; and their actions, in defence of their particular causes, often generate a political crisis, the result(s) of which can vary.

Social movements have three characteristics – they are involved in conflict with clearly defined opponents; they are linked by dense informal networks and they share a distinct collective identity (Della Porta and Diani 2006). Wilson emphasizes the organized and conscious character of social movements. A social movement is a conscious, collective, organized attempt to bring about or resist large-scale change in the social order by non-institutionalized means (Wilson 1973 as cited in Cannon 2003). Tilly emphasizes the challenge to power, the means it employs, and the difficulties of tracing social movements across time, while Byrne emphasizes the expressive character of social movements. Tarrow succinctly summarizes social movements as a form of contentious politics.

Contentious politics occurs when ordinary people, often in league with more influential citizens, join forces in confrontation with elites, authorities and opponents. Such confrontations go back to the dawn of history. But mounting, coordinating, and sustaining them against powerful opponents are the unique contribution of the social movement – an invention of the modern age and an accompaniment to the rise of the modern state Contentious politics is triggered when changing political opportunities and constraints create opportunities for social actors who lack resources on their own. They contend through known repertoires of contention and expand them by creating innovations at their margins. When backed by dense social networks and galvanized by culturally resonant, action-oriented symbols, contentious politics leads to sustained interaction with opponents. The result is the social movement (Tarrow 1998).

These descriptions give us an overview of the multifaceted interaction between civil society, NGOs and social movements. There are issues at the margins of these interactions, for example, are trade unions part of civil society or not? Are they NGOs or not? Similarly, there is no Chinese wall that separates NGOs from social movements, and in fact we may have a special type of NGO that exists as an expression of a social movement sometimes called a social movement organization.

The South Maharashtra Movement for Water Rights (SMM)

What we have called the South Maharashtra Movement for water rights (known as SMM hereafter) describes the contentious politics in south Maharashtra, a state in western India, around the issue of water rights and access, and reflects fully the complexity of the social interactions described earlier. The Maharashtra Rajya Dharangrasta va Prakalpagrasta Parishad (Maharashtra State Federation of the Dam and Project Affected, hereafter, the Parishad) and the Shetmajoor Kashtakari Shetkari Sanghatana (Agricultural Laborers and Toiling Peasants' Front, hereafter SKSS) are the two organizations representing a common movement in south Maharashtra. They share the same activists and leaders from the region, have common strategies and mobilize a common social base for both aims. The movement also shares a long history, with political groups like the Shramik Mukti Dal (Toilers' Emancipation Group, hereafter the SMD) and the Mukti Sangharsh Movement (Struggle for Emancipation, hereafter the MSM), organizations and movements like the Lok Vidnyan Sanghatana (People's Science Organization, hereafter the LVS) as well as individuals and people-oriented technologists like K. R. Datye. The formation of the movement ensured interaction between old and new social movements and concerns.

We shall trace the genesis and development of the movement – both as a movement of the dam and project affected, as well as for equitable access to water, especially Krishna waters, before we return to the role played by social movements and by contentious politics and confrontation.[1]

Genesis of the South Maharashtra Movement for Water

The Mukti Sangharsh Movement (MSM)

The establishment of the Mukti Sangharsh Movement (MSM) in 1983 was the first important organizational expression of the SMM. The origins of MSM can be traced to the famous 1982 textile workers strike in Mumbai, the capital of, Maharashtra. This historic, prolonged strike was much more an expression of accumulated anger simmering within the textile workers in Mumbai, rather than a calculated rational instrument of collective bargaining. It led to the dismantling of a sizeable section of the textile industry. Large numbers of textile workers, along with their leadership, who had returned to their villages, to sustain themselves during the prolonged strike, lost their jobs and were forced to stay on. Among them were a number of workers from the drought-prone, low-rainfall Khanapur taluka in Sangli district. Recurrent drought was common, as was drought relief work taken up under the Employment Guarantee Scheme (EGS). With their background of working class struggles, these workers became involved in organizing the EGS workers' rights, most of whom were marginal farmers and agricultural laborers. However, as MSM took up this task in its initial period, it also realized that drought was an important aspect of their life and that a strategy to eradicate drought had to be an important component of the emancipation it sought for its constituents.

Interaction with the People's Science Movement

The other strand from which the MSM drew its strength was its interaction with the People's Science Movement. LVS in Maharashtra represented what was a strong movement in the late 1970s and early 1980s. The MSM carried out a detailed survey of the situation along the Yerala river, a tri-butary of the Krishna that drains the western part of Khanapur taluka. Along with LVS activists, it also organized a Vidnyan Jatra (Science March or Fair) in 1984 to spread awareness about drought and how to deal with it. It helped bring together locals, scientists and technologists to collaborate on different ways of dealing with drought. This dialogue has been one of the strong points of the movement, and it continues to this day.

The 3D Committee for Drought Relief and Rehabilitation

Following the severe drought in Maharashtra in 1972, the Government of Maharashtra installed the "Drought Relief and Rehabilitation Committee," popularly known as the Dandekar-Deuskar-Deshmukh committee or the 3D committee, that continued its deliberations into the early 1980s. The Committee was headed by V. M. Dandekar, the renowned economist; V. R. Deuskar was a senior irrigation engineering expert and Datta Deshmukh of the Lal Nishan Party (Red Flag Party, LNP) who spearheaded the anti-drought movement sparked by the 1972 drought, was a self-made and self-taught farmer expert. The committee also included representatives from various political parties, and intellectuals from the famous Gokhale Institute of Economics and Politics, Pune (V. R. Dandekar was then the Director). Though it slowly languished and later became ineffective, it nevertheless played an important role in shaping the civil society's response to the severe drought in Maharashtra in 1972. It created a fertile environment for different left wing fronts to explore different strategies for organizing struggles around the drought. The LNP and their leader Datta Deshmukh not only led many such struggles, but also played a significant role in suggesting long-term solutions for dealing with the drought.

The report of the 3D committee and the various anti-drought struggles called for the state to rethink its irrigation policy. The committee had argued for an irrigation norm of 750 mm, which is equivalent to 7,500 cubic meters of water for a hectare. This was based on the assumption that irrigation would be provided for eight months, for the kharif and rabi seasons – a radical break from the earlier policy, since it did not assure water for the cultivation of crops like sugarcane, which are water intensive and require water throughout the year. The committee's recommendations however were not fully accepted (Phadke and Patankar 2006).

Through its association with the LVS and a group of pro-people experts from the Centre for Applied Systems Analysis in Development (CASAD), MSM also reviewed the 3D committee's recommendations. One of the important understandings that emerged was that, instead of totally depending on external irrigation, water should be used as a supplement to local water resource and watershed development. Detailed on-farm research was carried out by the MSM along with CASAD, as part of the Wasteland Integrated Research Program (WIRP), which showed that a minimum assurance of water and biomass input from wasteland areas could ensure livelihood needs even in drought years and helped evolve alternative norms of water requirements for livelihood needs. This spadework went a long way in the struggles that ensued. The interface with scientists and the systematic study that backed their demands was an important and distinctive aspect of the movement.

The MSM and its development was closely related to the return of the workers and the leading role in the movement taken by the activists of the Shramik Mukti Dal (SMD), a political group committed to grassroots action and radical transformation. The MSM worked mainly in Khanapur taluka of Sangli district, but towards the late 1980s, MSM activists joined two larger struggles – the struggle of the dam-affected and project-affected in the area, and the struggle for more equitable access to water, more specifically waters of the river Krishna.

The Struggles of the Dam-affected People — Dharangrasta Parishad[2]

The state of Maharashtra is credited with the largest number of dams in India. While the dominant narrative translates this into prosperity, it has often been at the cost of displacing large numbers of people. The dam-affected and the project-affected people in Maharashtra came together to form the Maharashtra Rajya Dharangrasta va Prakalpagrasta Shetkari Parishad (Maharashtra State Federation of the Dam and Project Affected, or the Parishad) because the large scale displacement of farmers largely remained unaddressed. Under the leadership of Dr Baba Adhav (a socialist leader of workers and toilers in the informal sector) and Datta Deshmukh, this organization became a major force to reckon with. Its consistent struggles had a role to play in the state government enacting the Maharashtra Rehabilitation of The Project-Affected Act in 1986. This Act, then one of the most progressive in India, provides, among other things, for 13 civic amenities (now increased to 18, ranging from houses and schools, to piped water supply and drainage) to be provided for the newly settled villages of dam-affected people. More importantly, it provides for a *minimum* of 2 acres of agricultural land in the command area to be provided to every dam affected family. The Parishad had made this a strong foundation for a progressive expansion of demands around water rights, as well as a basis for an alliance and negotiation between the dam beneficiaries and the dam-affected (Pune Support Group 2001).

The rehabilitation of the Koyna dam-affected families was the oldest and the worst, and the struggle around it continues even today. However, after the formation of the Parishad, rehabilitation performance has generally improved. During the 1990s, there was a fresh spurt in the struggles of the dam-affected people in Maharashtra, mainly related to the Bachawat Award of 1975. The Bachawat Award had specified the allocation of Krishna river waters between the states of Maharashtra, Karnataka and Andhra Pradesh (the concerned states sharing the Krishna river waters) and fixed the quotas that had to be utilized by them by impounding the Krishna waters by May 2000. Unutilized shares by then would become part of a common pool considered for reallocation in the next round of negotiations. The Government of Maharashtra therefore rushed to dam its share of the waters by May 2000. This led to a fresh bout of dam construction in the Krishna valley, and often without the necessary infrastructure to carry this water to the fields (Phadke 2000).

This led to an equally strong bout of resistance and struggle for proper rehabilitation from the Parishad. In south Maharashtra, in about 13 talukas (equivalent to blocks) in the districts of Sangli, Satara, Kolhapur and Solapur, the MSM merged with the Parishad and became a formidable force. It led the struggles of dam-affected people in Uchangi, Chitri, Urmodi, Warana, Wang-Marathwadi, Uttarmand and other dams in these districts. The resultant movement is characterized by its innovative demands and organizational strategies that have laid the foundation for the expansion of the water rights for the dam-affected.

Innovative Demands for Equitable Water Distribution

Though not directly related to the dam-affected, equitable water distribution has been a consistent demand by the SMM and is the basis of all their struggles, whether for the dam affected or the dam beneficiaries. It considers water as an important means of production, and has consistently struggled for equitable distribution of water, including to the landless. It has also made a simultaneous demand for redistribution of land. The insistence on minimum land for every family, including the landless, is related to this. The radical model promoted by the activists is now under negotiation as a pilot project in two talukas in the Krishna valley, and is discussed ahead in the chapter.

Pani Bhatta – Monthly Water Allowance

The movement has consistently built on the previous gains of the Dharangrasta Parishad. The demand for a monthly water allowance is one such demand. One of the activists' demands, to which the state government had agreed, is the provision of minimum 2 acres of *irrigated* land in the command areas for project-affected families. However, the delay in its implementation has meant a continuing loss for the people. The activists therefore demanded a pani bhatta – a monthly water stipend – as an interim compensatory measure, currently pegged at INR 600/month (US$15 approx) until the government can provide irrigated land as compensation. This has proved to be an important interim measure for the dam-affected, who have put their weight behind the demand. The demand has been conceded in areas like Warana and the struggle is underway to enforce elsewhere.

"First Rehabilitation, then Dam Construction"

The second important demand has been the demand for rehabilitation *before* construction of the dam. The movement has insisted on obtaining written orders to ensure appropriate rehabilitation of all dam oustees before any construction of the dam is allowed to proceed. In the case of the Urmodi dam, the Parishad did not allow the laying of the foundation stone until written orders had been issued for complete rehabilitation of every dam-affected person. This has become popular as the "Urmodi pattern." An important outcome of these struggles is the provision of 13 basic civic amenities for every rehabilitated family and a grant of INR 10,000 (around US$ 225) per rehabilitated family for the construction of a house. The militant mass mobilizations that have received wide support from the people and the local press despite police action, have sent a strong message to the government that this movement cannot be suppressed by the armed forces alone.

"Better Water Distribution from Old Dams, before Construction of New Dams"

Improved water distribution from existing dams, before the construction of new dams, is another demand rarely voiced by other movements. Often the emphasis is largely on rehabilitation issues, and this continues even after the dam is constructed. Not much is said about how the water is distributed. In a new dam-building frenzy, it is common to build a dam without even a canal network, release the water into the stream and leave it to the farmers to pick it up from there, effectively placing water rights in the hands of those who can afford to invest in pumps, pipes, channels and the like. The SMM has opposed this trend and promoted more equitable water distribution from existing dams. It demands allowing new dams to be constructed only after this has been done. Its insistence has forced the government to initiate work on canal networks on dams like the Urmodi and the Tembu lift scheme.

Creative Organizational Strategies: Joint Dam-affected and Dam Beneficiaries

Another major strength of the movement has been its creative and persistent organizational strategies. A highly innovative and effective strategy adopted by the movement has been to oppose the state by uniting the dam-affected and dam beneficiaries in common cause. This was made possible by another flank of the movement described in the next section that has worked for expanded access to water, completion of dam projects, adequate infrastructure and equitable access to water through the projects, and also combine these with the demands for complete rehabilitation of all the dam-affected, including their right to irrigated land. This cross fertilization has remarkably united two groups who would otherwise have been at loggerheads.

Mobilizing the Right to Information

Another important strategy has been to demand comprehensive information and propose concrete alternatives to the dam projects. In the case of the Uchangi dam in Ajra taluka of Kolhapur district, the activists, with inputs from experts, obtained critical information from the irrigation department and proposed a smaller dam-based scheme which would reduce submergence, without affecting the water available for irrigation. It successfully had the height of the dam substantially reduced, and effectively prevented the submergence of not even a single "gaothan" house (village habitat) in the main affected village. To create a viable alternative, the activists and the team of the Society for Promoting Participative Ecosystem Management (SOPPECOM) required topographical survey data. The State Water Resource Department refused to share this data, claiming it was confidential. Using the Right to Information, the activists obtained the data and submitted an alternative plan.

Consistent Follow-up

One of the movement's strengths is consistent and rigorous follow-up for complete implementation of demands. This seemingly routine matter has contributed to the many and real ground-level successes of this movement, that have made many long-drawn struggles incrementally worthwhile. The movement has developed an internal discipline amongst its cadres to enter a dialogue with specific information in hand, to ensure that every meeting or dialogue with the government officials is adequately recorded and written minutes obtained from the government. Simple though it may seem, these government documents have become important tools for follow-up and for actualizing the gains. For example, a struggle was launched in Satara in 2001, where the activists demanded that the government revisit some of its promises made, and gave it an ultimatum demanding that all the promises made and recorded should be met within a specified period of time.

Indefinite sit-ins–'thiyya andolans'

The consistent follow-ups have been combined with a novel form of indefinite sit-ins or "*thiyya andolans*."[3] People sit in front of government offices until they extract written promises from the government. In January of 2002 and 2003, respectively, over 3,000 dam-affected people launched an indefinite, round the clock "*thiyya andolan*" in front of the Satara District Magistrate's office for about 20 days. The dam-affected withdrew the agitation only after obtaining a written acceptance of their demands by the then chief minister.

Another massive indefinite sit-in was organized in Pune in 2004, when about 7,000 people from 13 talukas assembled to put forth their demands before the Maharashtra Krishna Valley Development Corporation (MKVDC). These demands were related to larger policy changes like the financial allocation of INR 30,000 million ($ 650 million) for drought eradication; change in the priority allocation of water from industry to agriculture;

equitable allocation of water in proportion to the population, and allocation of INR 5000 million (US$ 100 million) for rehabilitation of the dam-affected. This strategy combined the demands of the dam-affected and drought-affected dam would-be-beneficiaries. The sit-in was planned almost over a month with major campaigns launched in different districts, led by senior leaders of the movement in Maharashtra.

Evolving Specific Alternatives

Recognizing that many of their demands were long-term and may take time to be met, the activists took pains to present specific, practical alternatives, with the help of allies, pro-people experts and institutions. For example, the activists had demanded that the work on the Urmodi, Tembu, Uttarmand and Wang projects be completed and water be released to the talukas of Karad and Patan for the rabi (winter season) crop. The government had said that new projects could be taken up only if funds are available after completion of the older pending projects. Since it was apparent this would take some time, the activists demanded that water from the Koyna and Kanher dams be made available to these talukas during the interim period. These innovative strategies and demands cornered the government and they have not been able to contest the demands on any ground.

Similarly in the case of the Uchangi dam struggle, an alternative plan was proposed through systematic scientific enquiry and SOPPECOM's help. It proposed three smaller dams on the Tar-Ohal stream that could together impound 624 million cubic feet of water, with the potential to irrigate almost double the area marked in the government plan. The plan also proposed an equitable allocation of 3000 m^3 of water per household, and involved no displacement of habitation. The plan was well founded and the government could not contest it. However, the government accepted only one of the three proposed projects on the grounds of cost. Even so, they agreed to reduce the dam height by 2 meters, which substantially reduced the submergence and saved all habitats.

Struggles for Equitable Access to Water

The Pani Panchayat Movement

The wider struggle for equitable rights over water in the region grew from work done at the micro level by the Pani Panchayat in the 1970s and the Baliraja dam movement in the 1980s. For the first time, the Pani Panchayat movement conceptualized and grounded the idea of equitable access to water, irrespective of land holdings. Its key principles included, among other things, a ceiling on irrigation of 2.5 acres per household, and a refusal for allowing sugarcane cultivation as it is a water-intensive crop. The Pani Panchayat's most important contribution has been to separate water rights from land titles, and treat water rights as part of the farmers' right to livelihood. Although its experience was at a micro level, its impact was widespread and the SMM was one of the first organizations to engage with it in a meaningful manner.

The Baliraja Struggle

A little later in the early 1980s, when the MSM was formed and grew in strength, it initiated a struggle along the river Yerala, where sand was being excavated on a large-scale, for the profit of a few. The common resource that belonged to the villagers was thus being robbed by outsiders for profit. The excessive sand mining also led to drying up of wells in nearby farms. Even as MSM launched a struggle to stop the excavation, it also explored the possibility of building a dam on the river Yerala to benefit the villages of Balawadi and Tandulwadi. This led to the birth of the Baliraja dam.

The Baliraja dam symbolizes several things, as it was a first in many respects. The activists teamed up with pro-people scientists like K. R. Datye, Suhas Paranjape and Vilas Gore, who were then with the CASAD. It was one of those rare occasions, when diverse groups of people came together and decided on the site, proposed and built a dam – with proceeds from the sand excavations, funds from wider middle class public support and technical support from groups like CASAD. It was also the first time that people worked out a detailed plan for equitable, sustainable and rational use of water. The five year "on farm" study by CASAD, MSM and other local groups (1986–91) as part of the WIRP in different locations in Maharashtra, contributed immensely in developing norms for equitable distribution of water. This plan became a key component of a new strategy in drought eradication and sustainable agro-industrial development – the availability of reliable, external 3,000 cubic meters of water to every family, under control of the local people; a mixed crop pattern with a certain proportion between grains, vegetables, fruits, fodder, biomass for fuel and agro-industrial use (Phadke and Patankar 2006).

Movement Expands from Micro to Macro

This micro-level initiative of the MSM impacted the macro level all over south Maharashtra. The struggle over the Takari Lift Irrigation Scheme was soon followed by a movement for equitable water access in 13 drought-prone talukas in the Krishna basin in Maharashtra, resulting in the formation of the SKSS (discussed later in the chapter).

Takari Lift Irrigation Scheme

One of the early struggles took place around the Takari Lift Irrigation Scheme on the river Krishna to correct regional imbalances in water allocation. The argument was that almost all the Krishna water was being utilized in the immediate vicinity of the banks of the river, whereas the further, higher placed, drought-prone areas in the Krishna valley who needed it the most hardly received any water. The main demand was an equitable share of the Krishna waters for the drought-prone areas, even if it meant higher lifts. Another demand was that water allocation for a region should be in proportion to its population.

The Takari Lift Irrigation Scheme, that served a portion of the drought-prone plateau by lifting water, was announced by the government around the mid-1980s. The plan was to lift 4.6 TMC (thousand million cubic feet) of water to a height of 116 meters using 31 MW power, and convey it to the drought prone areas of the region. The total financial outlay for this scheme was INR 2,800 million (US$ 70 million). According to the government plan, only about 13,000 hectares would be irrigated, fully covering eight villages and partially covering 22 villages. The MSM consulted experts from CASAD and put forth a plan using its norm whereby all households would be equitably covered and supplied 3,000 cubic meters/household, irrespective of landholding. By this plan, they showed that it was possible to cover 60,000 ha across 60 villages. The struggle was launched in May 1989, with the demand for equitable water distribution of the Takari scheme according to their plan. A memorandum was signed by over 1,000 peasants from the drought-prone areas, and similar resolutions were passed in several gram panchayats (village bodies).

The persistent struggles launched in different locations finally forced the state government to yield partially to the MSM's demands and in October 1989, the government agreed to some demands. Though equitable access was not accepted as a principle for the entire project, the MSM was allowed to form village level Water User's Associations (WUAs). Earlier, the government insisted on a WUA that would serve only the land in the designated command area on a pilot basis, and would enter into an agreement with the Irrigation Department, so that they could utilize the quota for their village on an equitable basis. But the activists changed this to include villages outside the command areas – a major victory that allowed them to form WUAs in their own strongholds and scale up its demand for the right to water. For the first time, the state government allowed equitable water distribution as a principle. This victory was a turning point for the movement, and drew the attention of the media, political leaders and intellectuals and importantly, gave the masses confidence that they were on the right path.

Formation of Shetmajoor Kashtakari Shetkari Sanghatana (SKSS)

Inspired by the success of the Takari struggle, the MSM and prominent leftist leaders in the region came together to form the

Shetmajoor Kashtakari Shetkari Sanghatana or the Agricultural Labourers and Toiling Peasants' Front (hereafter referred to as the SKSS) in 1993 in a massive rally in Kini village in Kolhapur district, attended by over 20,000 people. The SKSS launched a much wider movement for equitable access to water. Every year in July, SKSS holds a rally to press for equitable access to water, including, equitable access to the water of river Krishna for different regions.

A series of struggles in 13 talukas in the drought-prone districts of Sangli, Satara, Kolhapur and Solapur were launched simultaneously to take forward the struggle for equitable access to water. Every new scheme in these areas was seen as an opportunity to push for the demand, making equitable water access a household issue. In fact, the parties who were elected in 1999 in Maharashtra had to include equitable distribution of water in proportion to the population as the first of the 51-point Common Minimum Programme (CMP), on the basis of which they sought votes from the people.

Tembu Lift Irrigation Scheme Modified for More Equitable Access

The Tembu Lift Irrigation Scheme provided an opportunity for the SKSS to propose a more equitable alternative. It is a huge lift irrigation scheme that aims to lift 22 TMC of water from the Krishna river basin in five stages to irrigate 79,000 ha of land in 173 villages across six talukas. The struggle was strongest in Atpadi taluka in Sangli district, one of the worst affected by drought and assessed to have the lowest per capita water availability in the entire rain shadow region. The SKSS demanded a restructuring of the scheme to allow it to serve a larger population based on the principle of equitable access to water.

In fact, the movement was strongest in the Atpadi taluka. Activists showed that it was possible to cover *all* households in the taluka, that is, about 22,000 families and allocating water of 5,000 cubic meters per family (Phadke and Patankar 2006). In the old plan, 21 villages were completely left out, and in the rest only farmers with land in the command area received the benefits of irrigation. The activists also demanded that the government bear the cost of equitable sharing of the waters. The government has finally agreed to restructure the Tembu scheme – currently only in Atpadi taluka – in line with the activists' alternative.

The Tasgaon Pattern

While the Atpadi farmers were busy organizing the restructuring of the Tembu scheme for equitable water distribution, similar initiatives were spreading in other talukas as well. Historically, Khanapur and Atpadi have been at the center stage of SKSS's water struggles in south Maharashtra, rather than Tasgaon. However the severity of the drought, the indecisiveness of the political leadership and SKSS's inspirational leadership changed the situation dramatically. Activists organized a series of big rallies, demanding that 40 drought-prone Tasgaon villages be included in the service area of the Takari and Tembu lift irrigation schemes. The Minister of Irrigation of the state accepted the demand in principle in writing. Activists from the 13 talukas who came there celebrated this remarkable victory of December 6, 2003 as Dushkal Bimod Din (Drought Eradication Day). This struggle came to be known as the Tasgaon pattern and was followed in other talukas as well.

Specific Restructuring Alternatives for Equitable Access

The key demand of the activists was equitable distribution of water. But to ensure that individuals at the grassroots would benefit, they nurtured the conviction that water, like land, is a productive resource and therefore everyone had a *right* to a minimum assurance of this resource. Water was recast as part of the right to livelihood, *irrespective of landholding*. This vision is a very inclusive one and therefore attracted huge numbers. It includes the landless, women, people outside the command areas, artisan groups and all hitherto water deprived and insecure sections of society. Of course, this is easier said than done, but the activists made consistent efforts to succeed. For example, in the Baliraja struggle, the activists demanded that women be given a separate right to at least a small quantity of water. Operationalizing this right has not been easy, as it meant challenging patriarchal relations within the household. Nevertheless, the activists raised this issue on all public for a which enabled diverse groups to unite and collaborate creatively.

It has also creatively used the space within the state machinery to its advantage. For example, during a severe drought in Sangli district, the government launched a drought relief program with food for work, fodder for cattle and the EGS. The activists demanded that drought relief work through schemes like the EGS and Food for Work programs include, completing canal work in the lift schemes and at older dams which would facilitate more equitable water allocation. It performed a systematic study and pointed out to the government that instead of the INR 8000 million that the government planned to spend simply on relief work for cattle camps or new tank construction, could instead be used to complete work on the Tembu scheme canal, costing only INR 2,500 million, which would provide water to about 200 villages in four drought-prone talukas.

Organizational Strategies: *Padayatras*, Sit-ins, *Chavani Andolans,* Signature Campaigns and Non-cooperation

The Baliraja struggle set a trend for including *padayatras* (journeys on foot) along the river and canals, through people's lands, to understand the impact of new projects and to assess the feasibility of alternatives. Although intensive and time consuming, they imply a thorough inspection of the physical structures and features of the land, often carried out with the help of experts. They serve as tools for building awareness, planning and involving the entire population in discussions, for example, before proposing alternatives for the Takari scheme, activists had launched a one-month campaign, including a 10-day *padayatra* along the proposed canal. This allowed a detailed investigation into the extent of submergence, and also the possibilities of extending the command areas of the Takari scheme to include villages in the Tasgaon, Atpadi and Man talukas.

Activists have used indefinite sit-ins, *rasta roko* (road blocks), *chavani andolans* (where cattle block the taluka center, for instance) and signature campaigns to further their cause. In Man taluka, an indefinite sit-in later culminated in a *rasta-roko*, in which men, women and cattle blocked all the main roads in the taluka to demand sharing of Takari waters.

On another occasion an indefinite *chavani andolan* fortified a kind of people's court, before which all election candidates were invited to clarify their position on the issue of drought and development.

In Atpadi, activists collected over 66,000 signatures in eight days, on a memorandum demanding equitable distribution of water from the Dhom and Ujani dams to all households, including the landless laborers. The signatures were submitted to the then Chief Minister, Sharad Pawar. Since this did not receive much attention from the government, the activists altered their strategy and got all the 56 village panchayats to send in their resolutions to the government. Within a month, the activists managed to muster written support from diverse organizations in the taluka, including, co-operative societies, workers' unions, ex-army-men's associations, as well teachers and students bodies, and had them send resolutions to the chief minister, supporting the original memorandum. As a result, the resolution supporting the demand was passed in the taluka panchayat. Thus, the campaign involved the entire population.

Though conventional leadership is somewhat equivocal on the effectiveness of non-co-operation, activists from drought-prone areas have used it to a dramatic effect, particularly in Tasgaon, to demand their due share in the waters of river Krishna. The state government responded by announcing that it would grant Atpadi taluka a share of water from the dam to be built at Urmodi. Since the construction of the Urmodi dam had not yet begun, the promise was just a weak attempt to deflate the movement. So in 1994, the activists responded by declaring non-payment of land revenue until the demand for their share of Krishna water was met. This was a politically challenging step, unprecedented in this area since independence as it challenged the very legitimacy of the state. This was followed by a special women's *jatha* (gathering) program.

Women not only participated in the *jatha*, but also conducted its affairs that included building awareness amongst the participants. During this program, too, women vowed not to pay land revenue till their demands were met.

Contestation: Key Foundation of the Movement

The gains achieved by the SMM are in many ways unparalleled. Every gain, since the Baliraja struggle, has contributed to the movement and helped it expand geographically as well as conceptually. One of its greatest achievements has been the ability to cross-fertilize the movements of the dam-affected and the drought-affected into a single larger movement with a common aim.

Importantly, the transition from the micro to the macro did not compromise on the demands at the micro level. When the MSM challenged the regional imbalances in water allocations in the Krishna river basin, it did not give up its demand for equitable allocation at the household level. In fact, it made the issue of equitable access an entry point for raising the issue in households. Similarly, in Atpadi and Tasgaon, the demand for taluka level water allocations did not compromise on similar allocations at the village and household level. This is precisely the strength of the movement. For most movements, it is a challenge to stick to micro demands as they expand their base. Of course, the movement has faced stiff opposition from its external allies as well as internally, while insisting on equity at different levels, including the affected dam. It is, however, the severity of the drought in the region that makes equitable access such a pressing need – once the movement leaders decided not to compromise on these issues, they received enormous support from the masses, especially the toiling and deprived sections that they supported.

At the same time, the activists have taken experts' help to ensure feasible alternatives. Their strength lies in a combination of factors, including innovative ways of making demands, organizational strategies and generating a mass base, including the landless, small peasants, women and the working class. This has resulted in numerous concrete, long-term gains in expanding and defending water rights, especially for the poor.

There are also many aspects of the movement not covered here, including alternative approaches such as a low external input, sustainable agriculture, and biomass-based agro-industrial options. The activists are acutely aware of the energy costs involved in the Takari and Tembu lift schemes. They are evolving a scheme that earmarks a portion of the water and land in the taluka for biomass production and so are ready to pay for the energy costs through a renewable energy equivalent.

All these are characteristics of a social movement that has notched significant gains. It is difficult to imagine an NGO achieving the same objectives, for example, by uniting the dam-affected and drought-affected on a common platform; they would typically avoid confrontation and radical street action. An NGO's milder tactics, aimed at achieving specific limited aims, would be unlikely to make much headway.

It is crucially a process of contentious politics by, as well as on behalf of, the otherwise powerless, deprived and disadvantaged sections. Much more than an NGO, it has an expressive energy and militancy that sustains it. The poor in the 13 talukas of south Maharashtra have come to see in the movement, an expression of themselves and their strength, a strength they would never have had without it; it is the foundation on which all its gains rest.

Notes

1. Most of what follows has been based on Omvedt and Patankar (1991), Phadke (2000) and Phadke and Patankar (2006).
2. Part of what follows is based on a note by the Pune Support Group on the Dharangrasta struggle (Pune Support Group 2001).
3. *Thiyya* is a wedge driven to make things fast and "*thiyya deun basne*" is used for obstinate refusal to budge.

References

Allwood, G. and K. Wadia. 2000. *Women and Politics in France, 1958–2000.* London: Routledge.

Cannon, S. 2003. "Social Inclusion, Social Movements, and the Characteristics of Late Modernity," Bowater School of Management and Marketing. Geelong, Deakin University, Australia.

Centre for Civil Society. 2004. "What is Civil Society?," http://www.lse.ac.uk/collections/CCS/what_is_civil_society.htm (accessed on November 10, 2013).

Della Porta, D. and M. Diani. 2006. *Social Movements: An Introduction.* Oxford: Blackwell.

Hegel, G. W. F. 1820. *Philosophy of Right,* http://www.marxists.org/reference/archive/hegel/works/pr/index.htm (accessed on November 10, 2013).

Iriye, Akira. 2002. *Global Community: The Role of International Organizations in the Making of the Contemporary World.* Berkeley: University of California Press.

Phadke, A. 2000. "'Dam-Oustees' Movement in South Maharashtra," *Economic and Political Weekly,* 35(47): 4084–86.

Phadke, A. and B. Patankar. 2006. "Asserting the Rights of the Toiling Peasantry for Water Use: The Movement of the Dam Oustees and the Drought Affected Toilers in South Maharashtra," in P. P. Mollinga, A. Dixit and K. Athukorala (eds), *Integrated Water Resource Management: Global Theory, Emerging Practice and Local Needs.* New Delhi: Sage.

Prodi, Roman and Neil Kinnock. 2001. "The Commission and Non-governmental Organisations: Building a Stronger Partnership." European Commission Discussion paper, European Commission. Available online at http://ec.europa.eu/civil_society/ngo/docs/communication_en.pdf (accessed on November 10, 2013).

Tarrow, S. 1998. *Power In Movement: Social Movements and Contentious Politics.* Cambridge: Cambridge University Press.

Willets, P. 2002. "What is a non-governmental organization?," in UNESCO Encyclopaedia of Life Support Systems, pp. 3–4, City University, London. Available online at http://www.staff.city.ac.uk/p.willetts/CS-NTWKS/NGO-ART.HTM (accessed on October 27, 2013).

Wilson, B. R. 1973. *Magic and Millennium.* London: Heinemann.

11
Lessons from Plachimada
From Anti-Coca-Cola Agitation to Democratization of Water

C. R. Bijoy

Overview

Kerala's Plachimada struggle raises issues of the destruction of groundwater by a recalcitrant global multinational and the resultant threat to livelihood, particularly of adivasis (scheduled tribes/ indigenous people) in this case, who continue to be the worst victims of the corporate-state nexus. The persistent struggle by the local inhabitants, and the eventual response from the political, administrative and judicial system that oversee governance, have exposed inadequacies in governance specifically, those relating to water.

This article provides a narrative on the struggle and the way the governance structure and system responded, with reference to water and governance rights. The state, and implicitly the political democracy as we know it, meekly surrendered to the interest and dictates of corporate capital, despite a perceptible shift in favor of the struggle. There was also conflict of jurisdiction of decision-making power and misuse of laws and administration, the instruments of governance. The Plachimada struggle calls for communities to recover their rights over the resources belonging to or affecting the whole of a community.

While pointing out the changing nature of the concept and practice of the right to water, this article critically outlines the conflict between the constitutional provisions and statutory laws, most of which are a colonial legacy, that govern water and its use. This portends an intensification of the conflict between the state and the people. The case laws[1] arising from the judiciary also often contribute to perpetuating the conflict. However, the Constitution of India provides sufficient scope for the emergence of a democratic participatory governance system at its fundamental level – the community – especially in the case of groundwater and minor water bodies, the fundamental base on which the structure of water governance could be built upon. But are the people of Kerala ready to take forward Plachimada's gains?

The Beginning

The struggle over water rights in Kerala's Plachimada has been a long drawn out one, continuing since 2002, against the global giant, Coca-Cola, which polluted and depleted the groundwater. The Kerala state assembly unanimously passed the Coca-Cola Victims' Relief and Compensation Claims Special Tribunal Bill[2] on February 24, 2011[3] to extract compensation from the company for the losses caused by it. The Bill was sent to the President of India for approval.[4]

Earlier, on January 3, 2006, the Industries Department, the Kerala State Pollution Control Board (KSPCB), Perumatty Panchayat (local self-government) and Coca-Cola had met to consider Coca-Cola's proposal of November 28, 2005 to the Minister for Industries to relocate the plant from Plachimada, provided the government offered appropriate compensation. Nothing conclusive came out of this. The Plachimada plant was shut down in March 2004 as a result of the people's sustained struggle. Coca-Cola also reiterated that it will not sell its idling factory in Plachimada.[5] The state government constituted a High Power Committee in 2009 "to determine the extent and nature of loss resulting from the working of Coca-Cola company at Plachimada." The Committee submitted its report on March 22, 2010.[6] It severely indicted the company for polluting and depleting groundwater, violating as many as nine laws, and causing an estimated damage of at least INR 216.16 crore. Among other things, the Committee recommended setting up a claims tribunal to settle compensation claims.[7] The state government approved the recommendations and observations of the Committee and opted to constitute a tribunal to try Coca-Cola.[8] In the meanwhile, the Parliament enacted The National Green Tribunal Act 2010 for the "effective and expeditious disposal of cases relating to environmental protection and conservation of forests and other natural resources including enforcement of any legal right relating to environment and giving relief and compensation to persons and property" falling within the purview of the Water (Prevention and Control of Pollution) Act, 1974; the Water (Prevention and Control of Pollution) Act 1980; the Forest (Conservation) Act, 1980; the Air (Prevention and Control of Pollution) Act 1981; the Environment (Protection) Act, 1986; the Public Liability Insurance Act, 1991 and the Biological Diversity Act, 2002. However, as this law also stipulates that only cases where the cause of grievance is less than five years old is admissible in the Tribunal, the Plachimada case did not fall within its purview as the cause of grievance occurred well beyond the five year limitation period.

The demand for more powers to gram sabha (village assembly) in the control and management of groundwater has been a persistent one.[9] However, this was not part of the terms of reference of the High Power Committee which was limited to study the extent and nature of losses incurred by the Plachimada plant.[10] This central issue remains to be addressed.

Plachimada is in Perumatty Panchayat, one of the six panchayats in Chittoor block of Palakkad district, the rice bowl of Kerala.[11] The population in these seven worst affected colonies consists of about 30–40 percent tribals, 10 percent low caste Dalits[12] and other communities. The majority of the people are marginal farmers and agricultural laborers. Coconut and paddy are the main crops. The Ezhavas, a few Muslims and Other Backward Classes (OBCs) own almost all the cultivable land, with most of the Scheduled Tribes (STs) and Scheduled Castes (SCs) owning land ranging from 1 to 10 cents[13] used for housing. The STs, followed by SCs and most of the Muslims constitute the very poor, with agricultural wages averaging INR 80 for men and INR 40 for women for only about 120 days per year.

The main surface water source of the area, which falls in the rain shadow region of the Western Ghats, is groundwater and canal irrigation from the tanks and ponds, the Moolathara dam and the Kambalathara Lake. The left bank canal of the Moolathara irrigation project provides water for irrigation to the eastern half of Perumatty Panchayat. Of the 304 large tanks in Perumatty Panchayat, 291 are private and 13 are public tanks.[14] The Chittoorpuzha irrigation project of the Parambikulam-Aliyar Project (PAP), a Tamil Nadu–Kerala interstate project,[15] provides canal irrigation to the western part of the Panchayat. There are 2,804 public wells in the Panchayat and 838 irrigation wells. The 1,473 private dug wells and 508 borewells cater to domestic needs.[16]

Hindustan Coca-Cola Beverages Private Limited (HCCB)[17] acquired 34.64 acres in 1998, mostly paddy fields, in Ward 9 of Perumatty Panchayat to establish its plant at Plachimada in Moolathara village. The plant, located on the Palakkad-Meenakshipuram-Pollachi road, has proximity to many water sources.[18]

HCCB, after obtaining a licence[19] from Perumatty Panchayat on January 25, 2000, commissioned the plant in March 2000 to produce brands such as Coca-Cola, Sprite, Fanta, Kinley Soda, Limca, Thums-Up and Maaza, with three lines of production namely Returnable Glass Bottle Line, PET Line and Maaza Line. Permission was granted by the KSPCB to produce 561,000 liters of soft drink per day, requiring 2,131,800 liters at 3.8 liters of water per liter of soft drink. The HCCB employed about 130 permanent workers and 250 contract laborers. As per the factory records 500,000 liters of groundwater was withdrawn daily through five borewells and two dug wells.[20]

Adivasis Launch Struggle Against Coca-Cola[21]

Within a year of commencing the operation of the HCCB plant, the water table receded around the plant and water quality deteriorated soon, spreading to a 1–1.5 km radius of the plant. Agricultural operations, the lifeline of most people, were hit. Water became unfit for human consumption and domestic use.

HCCB initially sold the foul smelling slurry and sludge waste as fertilizer to unsuspecting farmers, later gave it away "free," and still later, surreptitiously dumped around. What was welcomed as "development" of this backward region turned out instead to be a threat to their very survival.

The Coca-Cola Virudha Janakeeya Samara Samithy (Anti-Coca-Cola Peoples' Struggle Committee) launched its struggle on April 22, 2002, with a blockade of the plant, demanding that it be shut down, and HCCB held criminally liable for pollution and depletion of groundwater, devastation of livelihood resources and endangering lives. The HCCB's attempt to get the High Court to intervene to prevent the continued protest failed. However they succeeded in ensuring police protection to the plant. The police, bending backwards, intimidated, harassed, arrested and foisted false cases against the protesters. These demonstrated in no uncertain terms not only the servility but complicity of the government and state to corporate capital, especially foreign investment, even when it posed a grave threat to its citizens. The struggle however remains peaceful and sustained.

As early as March 4, 2002, the Chennai-based Sargam Metal Laboratories concluded that the water in the area was unfit for human consumption and domestic use. The government primary health center corroborated this, based on the analysis by the state government's Regional Analytical Laboratory and asked the Panchayat in May 2003 to warn the public. In July 2003, the British Broadcasting Corporation's (BBC) Radio 4 "Face the Facts" program reported high levels of carcinogens, such as cadmium and lead, in the wastes dumped as "fertilizer" by HCCB in the farmlands, besides nickel, chromium and zinc in lower levels. The KSPCB confirmed this in September 2003, reporting more than double of what the BBC reported and four times the legally prescribed norm.[22] The KSPCB ordered HCCB to stop supplying waste, to recover all the waste transported outside and store it safely within the plant site. This was followed by the visit of the Supreme Court Monitoring Committee on Hazardous Wastes[23] to Plachimada on August 12, 2004. It indicted HCCB for the unauthorized disposal of hazardous sludge, non-disclosure of the source of toxic heavy metals and ordered HCCB to provide drinking water to all persons in the vicinity of the plant. Further, the report of the "Joint Committee on Pesticide Residues in and Safety Standards for Soft Drinks, Fruit Juice and Other Beverages" of the Parliament, constituted following the expose by the Centre for Science and Environment (CSE), New Delhi, on pesticide content in soft drinks, including Coca-Cola, stated that, "sludge analysis revealed cadmium content up to 338.8 mg/kg" when the maximum permissible limit set by the Hazardous Wastes (Management and Handling) Amendment Rules, 2000 is 50 mg/kg. The Central Pollution Control Board (CPCB) advised the KSPCB, "to direct the company to dispose the effluent treatment sludge as per the Hazardous Waste Rules." *Outlook* news magazine carried out a water analysis in April 2005 from a well in Vijayanagar Colony reporting that the water did "not meet the requirements for most of the parameters tested for potability as per ISO 10500 specifications set by the Bureau of Indian Standards" and considered the water to be dangerous if consumed.[24] The Hazards Centre, New Delhi and People's Science Institute, Dehradun, too, in a report[25] released in June 2006 reported high levels of toxic metals in it, concluding that the total natural sources around the plant were contaminated.

The Battle over Laws: Evading Justice

The Perumatty Panchayat issued a show cause notice in April 2003 for cancellation of the licence issued to HCCB under the Kerala Panchayati Raj Act, alleging depletion of groundwater resulting in drought in Plachimada, environmental issues and the agitation since a year by adivasis, which was challenged by HCCB.[26] The panchayat cancelled the licence nevertheless in May, asking that production be stopped. The Local Self-Government Department (LSD), when approached by HCCB, as advised by the High Court, stayed the decision of the Panchayat on the grounds that the Panchayat had exceeded its powers as conferred under the Kerala Panchayati Raj Act and later ordered the Panchayat in October to constitute an expert team to investigate the allegations. The High Court confirmed[27] that the Panchayat was within its powers to cancel the licence. The court later directed HCCB, who had challenged the cancellation of the licence as arbitrary, prejudiced and not based on facts or substantiated scientifically, to close all the borewells and to stop extracting groundwater beyond what was required for irrigating 34 acres of land. The court directed the government to carry out investigations into all allegations against HCCB. The court in its December 2003 order significantly also held that groundwater was public property held in trust by a government, and overexploitation of resources to the detriment of the public by a private party was not permissible. The action of the Panchayat was justified in consideration of public welfare above private rights, despite the rights of the property owner to extract the groundwater. However, the Division Bench[28] stayed the Single Bench decision.

In January 2004, a High Court-appointed expert committee was asked to monitor the water consumption, along with the Panchayat, who was also asked not to take any steps that would lead to the closure of the plant. Palakkad district, suffering from drought since 1998 along with 13 other districts, was declared drought-hit. The government banned groundwater use by HCCB until mid-June, when the monsoon was expected to set in. The High Court, while directing the Panchayat to renew the licence, also directed HCCB to look out for an alternative source of water, upholding the ban order of the government regarding drawing of groundwater. With this, the Plachimada plant ceased operations from March 9, 2004. The Panchayat rejected HCCB's application for a five-year renewal of licence that came up in February, and the court stayed the decision of the state government to overturn the Panchayat's decision. The KSPCB also rejected the HCCB's application of September 2004 to renew their "consent to operate" from 2005, for not providing details of the source of cadmium in the sludge at a concentration of 200 to 300 mg per kg of sludge, which is 400 to 600 percent above the tolerance/permissible limit,[29] besides failing to comply with the Supreme Court Monitoring Committee order to supply piped drinking water to the affected families, and installation of Reverse Osmosis Systems or any other more efficient system for better treatment of effluents.

The protests continued to exert pressure, garnering wider support across the state. The factory was blockaded on the 1000th day of the struggle on January 15, 2005, declaring that the factory would not be allowed to reopen. The "Final Report on Investigations on the Extraction of Groundwater by Hindustan Coca-Cola Beverages Private Limited (HCCB) at Plachimada" by the High Court-appointed Investigation Team[30] restricted its withdrawal upto 5 lakh liters per day, and lesser ac-cording to the extent of rainfall, which the court approved despite allegations of flaws in the Report. However, both the court and the Investigation Team did not deem fit to investigate the quality of water, which indeed was the primary problem for the people of Plachimada. The court observed that in the absence of a prohibitory statute, a person has the right to extract water from his property. Though the Kerala Groundwater (Control and Regulation) Act was passed in 2002 and notified in 2003, the rules were yet to be notified and so did not enter the arguments in the court. The High Court ordered the Panchayat to issue the licence to HCCB, which was met with protests across the state, while the people of Plachimada dared HCCB to operate the plant. The Panchayat rejected the application of HCCB, citing lack of clearance from the KSPCB, which the court disapproved. In June 2004, the court ordered the Panchayat to issue the licence within a week of application by HCCB, who could deem to have received the licence even if the Panchayat did not actually issue the licence, provided the company received all mandatory clearances. Faced with this inevitability, the Panchayat issued the licence, imposing various conditions that were not accepted by HCCB, which attempted to commence operation of the plant on August 8, 2005. The KSPCB once again rejected HCCB's application citing incompleteness of the application and non-fulfilment of the Supreme Court Monitoring Committee orders, and ordered the shutdown of the plant on August 19. In a turnaround, both the Panchayat and the state government, moved the Supreme Court in May 2005 and September 2005, respectively, challenging the High Court orders. With the Kerala Groundwater (Control and Regulation) Act getting activated with the notification of the rules in March 2004, Chittoor Block was declared a "notified area" by the Water Resource Department in November 2005.

In January 2006 the Panchayat reissued the licence with conditions, including non-extraction of water from the block, citing its "notified area" status. The KSPCB again rejected the application, preventing HCCB from commencing operations. With the Left Democratic Front assuming power in April – May, the government, based on the CSE's second report of August 2006 on pesticides in soft drinks, banned the manufacture and sale of Coca-Cola and Pepsi-Cola in the state in the interest of public health under the Prevention of Food Adulteration Act, 1954. This was struck down by the High Court as going beyond the powers of the state.

Further, the Kerala Groundwater Authority decided in March 2007 to carry out a detailed study based on the submission of the Struggle Committee, which demanded action against HCCB under the various provisions of the following:

(a) Kerala Groundwater (Control and Regulation) Act, 2002;
(b) The Water (Prevention and Control of Pollution) Act, 1974;
(c) Environment Protection Act, 1986, amended in 1988;
(d) Hazardous Waste (Management and Handling) Rules, 1989 and as amended in 2003;
(e) SC/ST (Prevention of Atrocities) Act, 1989
(f) relevant sections of the Indian Penal Code and Criminal Procedure Code.

Under the Kerala Groundwater (Control and Regulation) Act, 2002, with the declaration of Chittoor Block as a "notified area," all existing wells are to be registered, and the geology and groundwater conditions were to be examined, taking into consideration the purpose for which water was used, the other existing users of that locality, the rate of recharge of the area of influence of the well, the quality of groundwater on location, the long-term water levels of wells, and other relevant factors.

Further, it was demanded that all industrial production in the "notified areas" be banned or grant of permission denied for use of groundwater to those industries where

(a) water constitutes the major raw material or the primary content of a product;
(b) where the product falls under the category of a non-essential item (such as soft drinks in this case); specifically the application of HCCB for registering the wells be rejected and

comprehensive analysis of the groundwater in the affected area be carried out in terms of quality, identify the source of particularly the toxic contaminants, especially cadmium and lead, initiate criminal proceedings and hold HCCB liable on the basis of the polluter pays principle.

What is significant is the extreme reluctance of the political and administrative machinery to proceed as mentioned earlier, despite specific legal provisions, and instead diverting attention to peripheral issues such as the powers of the Panchayat versus those of the state, licensing procedures and fulfilment of requirements, rather than the substantive issues of endangering lives, destruction of the environment and criminal liability of HCCB. Only on September 27, 2007, the KSPCB finally issued notice to Coca-Cola for proceeding under Sections 43 and 47 of the Water Act, 1974, which referred to penalties for disposal of polluting matter by companies. Significantly, the judiciary, in this instance, took great care to stay clear of larger public interest, despite the collusive role of the government, the political executive and administration. By refusing to initiate the "judicial activism," that this case naturally called for, the judiciary also chose to ignore the intent of the laws, preferring to concentrate on procedural technicalities, rather than the substantive issues of threat to life and the environment.

From Anti-Coca-Cola to Water Rights: The Struggle Beyond

Kerala's Plachimada struggle carved out a distinctive place for itself among issue-based democratic struggles in India, for taking on Coca-Cola, the most well-known global brand and symbol of imperialism by the United States (US), that too by adivasis – the most marginalized people, in a sustained and persistent manner. The struggle captured the imagination of people within the state and even internationally, especially amongst the university campuses in the US[31] and UK. The struggle forced the Coca-Cola plant to shut down.

The adivasis and Dalits of Plachimada who are the poorest, mostly landless and illiterate agricultural workers, remain at the forefront of the struggle, rather than the peasants whose farms have been affected. The farming households, owning mostly over an acre, are better educated and are also involved in non-agricultural activities including trade, business and salaried employment. Unlike them, the adivasis and Dalits face intense water insecurity, are vulnerable, and their very survival is threatened by the Coca-Cola plant. The farming households adopted strategies to augment irrigation potential.[32] Economically, socially and politically, an insignificant village, Plachimada's battle against the multinational giant was interpreted as the "mischief of a few politically motivated persons" with "a handful of persons" at the HCCB plant making "false, motivated allegations." It irked the local power elite and those who had hoped for better opportunities arising from the HCCB plant and associated peripheral development. Being a protest by adivasis, lacking support from political parties and movements, the success of the state-wide popular adivasi struggle in 2001 for land,[33] and their support to Plachimada emboldened the people. The Adivasi Samrakshana Sangham, a local organization, constituted the "Coca-Cola Virudha Janakeeya Samara Samithy" (the Struggle Committee, in short) and commenced a round-the-clock protest outside the plant, besides other forms of struggle, deploying a simple spirit of resistance, determination and dignity. The Perumatty Panchayat was unhelpful, political parties antagonistic, and the media favored HCCB and the opponents of the struggle.

In the initial months, the major local political parties such as the Janata Dal, Communist Party of India (Marxist), the Bharatiya Janata Party and the Congress opposed the struggle. The HCCB, while harping on their environmental and socially responsible approach, belittled the struggle as "politically motivated," hinting obtusely to the involvement of Maoists, so that the state could take particular pride in crushing the struggle. Will an internationally acknowledged brand be so callous as to harm their reputation with such blatant acts? Initial skepticism soon gave way, as word spread of the genuineness of the struggle. Diverse sections from the extreme right to the extreme left, from neo-Gandhians to environmentalists, and from children to peasants from across the state, soon converged on Plachimada in support, organizing rallies, public meetings, blockades, sit-ins and *dharna*s. The supporters, a loose motley crowd, constituted the Plachimada Struggle Solidarity Committee, drawing from over 30 organizations across the state. The pro-HCCB media turned around from grudgingly reporting the struggle to enthusiastic coverage. Acknowledging the groundwater problem but shrugging off any guilt, the HCCB offered drinking water "magnanimously" to the affected villages, along with rainwater harvesting within the plant and outside in a farcical attempt to demonstrate their "corporate social responsibility," but in fact this was a public relations exercise to counter "ill-motivated and unfounded allegations." The villages rejected HCCB's offer, instead insisting on the Panchayat's responsibility in providing water.

The Communist Party of India (Marxist), the then leading opposition party in the state (and later went on to head the coalition government in the state till mid-2011), organized protests against HCCB, followed by the Janata Dal which controlled the Perumatty Panchayat. This, in effect, marginalized the local party leaders who had opposed the struggle. Except for the then ruling Congress (I), all other small and larger political formations such as Communist Party of India, Socialist Unity Center of India, Communist Party of India-ML (Red Flag), Communist Party of India-ML (Liberation), Samajwadi Jana Parishad and a host of others came out in support of the struggle. What started off as a local people's struggle, went mainstream as part of the dominant political discourse. Diverse organizations in and outside the state, such as the Adivasi Gothra Maha Sabha, All India Youth Federation, Democratic Youth Federation of India, Yuva Janata Dal, Solidarity Youth Movement, Gandhi Yuva Mandalam, Yuvajana Vedi, Peoples Union for Civil Liberties, Lohia Vichar Vedi, National Alliance of People's Movements, Kerala Sasthra Sahitya Parishath, Kerala Swathanthra Matsya Thozhilali Federation, Desiya Karshaka Sangham, Students Christian Movement, Poraattam, Small Traders Association, Pattikajathi-Pattikavarga Munnani (Scheduled Caste-Scheduled Tribe Front), Palakkad Munnottu, Swadeshi Jagran Manch and many other groups actively supported the struggle.

The Left Democratic Front assumed power in the 2006 assembly election, and promised to address issues that generated the struggle, assess the impact, address the health problems of the public, withdraw all criminal cases on the activists and participants in the struggle (orders were issued and cases withdrawn) and explore criminal proceedings against HCCB. However, decisive action was not forthcoming.

With Chittoor Block being declared a "notified area" under the Kerala Groundwater (Control and Regulation) Act, 2002, all existing wells in the area had to be registered, the geology and groundwater condition examined, and the application for registration of wells was to be considered on the basis of the purpose for which water is used, the other existing users of that locality, the rate of recharge of the area of influence of the well, the quality of groundwater in the location, the long-term water level of the well and other relevant factors. Despite sufficient evidence on pollution, no action against HCCB has been taken by the KSPCB and the Kerala Groundwater Authority despite clear violations of various provisions of the Water (Prevention and Control of Pollution) Act, 1974; the Environment (Protection) Act, 1986, and Hazardous Waste Management and Handling Rules, 1989 and as amended in 2003.

What has emerged substantively from the political, administrative and judicial response are (a) the nature of power of the local panchayat, regarding its responsibility towards public welfare and the jurisdictional division of power between the Panchayat and the state government in protecting and regulating natural resources and (b) procedural correctness in matters related to licensing for production and operation of the plant. But the struggle exposed the weakness of the political democracy which dodged its responsibility in taking a political decision on the basis of existing laws, and the judiciary chose to busy itself with peripheral issues, both indicating subservience to global capital.

The criminal liability of HCCB on the issues of quality (pollution) and quantity (water depletion) are yet to be addressed. Nor is there clarity on the fundamental issues thrown up by the struggle, as to who has the primary decision-making rights over water (groundwater in this case) – the people, the elected Panchayat, the state or the center; what are the primary uses of water – domestic, irrigation, or industrial; and who decides and sets the priority over groundwater use; and what should get priority – water for survival or water for profit. These issues are yet to be taken forward by a political process.

Democratization of Water and Governance Rights

Globally, under a neo-liberal economic dispensation, the "right to water" is quietly being reinterpreted to mean "right of access to water," which necessarily requires the "citizen" to be reduced to a mere "consumer" for the emerging high-growth "water market." The traditional and customary notions of natural rights to water, subject to the realities of social relationships and social distancing, were superimposed with state control and guided by the Constitution, to promote natural justice and equity. Instead, this in effect relegated ownership and control of water to the state, the propertied (landed) class and the market/business (capital).

The new market regime further disenfranchises vast sections of the marginalized and powerless. With water-intensive urbanization, industrialization and agriculture, modern water management systems, with their structures and technologies, the runaway decline in quantity and quality of water, and the escalating demand for water by high-growth centers, the focus is on development of water resources and supply-efficiency to meet different demands, rather than an overhaul of the water governance system through a democratic process. This, in turn, increases and intensifies the conflict around water. The citizen is a consumer or a potential consumer, and not in fact a right holder. A review and suggestion for an alternative water governance system is attempted here.

The National Water Policy, 2002 advocates a participatory approach to management of water resources. The scope of participation is determined by the following:

(a) the constitutional framework;
(b) the legislative process;
(c) the governance approach;
(d) conflicts;
(e) scope of resolution.

Article 21 of the Constitution of India on Fundamental Rights confers the right to life, liberally interpreted by the Supreme Court as not merely of survival or animal existence, but a life with human dignity and the right to livelihood, but according to the due process of law. The Directive Principles of State Policy recognize the principle of equal access to the material resources of the community though "shall not be enforceable by any court, but the principles therein laid down are nevertheless fundamental in the governance of the country and it shall be the duty of the State to apply these principles in making laws" (Article 37) and hence non-justiciable. Article 39 (b) mandates that "the State shall, in particular, direct its policy towards securing that the ownership and control of the material resources of the community are so distributed as best to subserve the common good." Article 15(2) explicitly refers to equal access to water with regard to "the use of wells, tanks, bathing ghats." Article 51-A (g) says it is the fundamental duty of every citizen of India "to protect and improve the natural environment, including forests, lakes, rivers, wildlife and to have compassion for living creatures."

The constitutional intent is to be translated through legislation with legislative powers distributed between the Parliament and the Legislative Assemblies of the different States. Article 245 to 255 on legislative relations between the Parliament and the State provides for law-making powers to the Parliament for the whole or any part of the country, and a legislature of the State for the whole or any part of the State. The topics of legislation on which the Parliament has exclusive powers to make laws, are set out in List I (Union List) while that for the State legislatures are in List II (State List) of the Seventh Schedule. List III is the "Concurrent List" of topics over which both the Parliament and State legislatures can make laws. Entry 56 in the Union List pertains to "regulation and development of inter-state rivers and river valleys, to the extent to which such regulation and development, under the control of the Union, is declared by Parliament by law, to be expedient in the public interest." In the exercise of this power, the Parliament enacted the River Boards Act, 1956, which has remained largely inoperative. Resolution of disputes between states in the sharing of river waters or river valleys, their use, distribution or control finds place in Article 262, which envisages the creation of an exclusive tribunal for the purpose by Parliament. Accordingly, the Parliament enacted the Inter-State Water Disputes Act, 1956.

The Parliament can, under Article 252, make laws even on topics on which it has no powers, provided the legislatures of two or more states resolve that Parliament should make such a law. Thus, the Water (Prevention and Control of Pollution) Act, 1974, a law on a topic relating to Entries 6 ("public health and sanitation; hospitals and dispensaries") and 17 ("water, that is to say, water supplies, irrigation and canals, drainage and embankments, water storage and water power subject to the provisions of Entry 56 of List I") of the State List was passed. The inadequate and unsatisfactory nature of local laws provided the justification for a central law to tackle the growing problem of pollution of rivers and streams. Another significant central legislation is the Environment (Protection) Act 1986, under which the Ministry of Environment and Forests laid down the regulatory framework for activities along the vast coastline of India.[34] The Government of India also set up the Central Groundwater Authority by a notification under this statute, to regulate the existing indiscriminate use of groundwater in various parts of the country.

Different states have enacted a number of statutes on various aspects of the control, regulation and distribution of water, such as laws that relate to canals and irrigation,[35] use of water sources,[36] water sewerage and drainage,[37] and groundwater.[38] The aspects of collection of taxes and cesses on the use of water are also covered by legislations enacted by both Parliament[39] and state legislatures.

The approach to legislation follows the colonial command and control governance model, a colonial legacy, a model for the subjugation of people, their territories and resources. This framework for governance, therefore, increases the resource-related conflicts between the state and people. A centralized regulatory authority with the power to grant licences, generates bureaucratic mechanisms in the chain of command that facilitate arbitrary decisions, susceptible and vulnerable to influence, red-tapism and corruption. This model, largely impervious to the democratic aspirations of participatory decision-making, transparency and accountability, runs contrary to the objective of decentralized community-based control of water recognized in the Constitution in the 73rd Amendment Act, 1992, by which the Panchayat as an institution of self-government, comprised elected members, was formally recognized.

Article 40 under Part IV of the Directive Principles of State Policy insists that the state shall take steps to organize village Panchayats and endow them with such powers and authority as may be necessary to enable them to function as units of self-government. Part IX of the Constitution deals with Panchayats and Article 243B mandates that "there shall be constituted in every state, panchayats at the village, intermediate and district levels in accordance with the provisions of this part." The provisions provide for administrative and legislative powers of the Panchayat. Under Article 243G, the legislature of a state can make a law to empower the Panchayat to prepare plans "for economic development and social justice," apart from other matters specified in the Eleventh Schedule to the Constitution, which include minor irrigation, water management and watershed development (Entry 3), fisheries (Entry 5), drinking water (Entry 11), waterways (Entry 13), health and sanitation (Entry 23), public distribution system (Entry 28) and maintenance of community assets (Entry 29). But this scheme of decentralization of responsibilities is yet to be effectively operationalized and the authoritarian bureaucratic command and control model dominates and conflicts with the Panchayat system. Moreover, Article 243 on Panchayats under Part IX gives pre-eminence to the state, rather than village Panchayats as envisaged under Article 40.

The Panchayats (Extension to the Scheduled Areas) Act, 1996, while extending the provisions of Part IX of the Constitution relating to the Panchayats to the Scheduled Areas as referred to in Clause (1) of Article 244, also modifies Part IX by directing that the state laws shall be in consonance with traditional management practices of community resources (Section 4.a); every gram sabha (village assembly) shall be competent to safeguard and preserve the community resources (Section 4.d); approve the plans, programes and projects (Section 4.e.i); identify and select beneficiaries (Section 4e.ii); and certify utilization of funds (Section 4.f); plan and manage minor water bodies (Section 4.j); and control local plans and resources (Section 4.m.vii). Further, the panchayats at the higher level are not to assume the powers and authority of any panchayat at the lower level or of the gram sabha (Section 4.n). This introduces the pre-eminence of the village panchayat, especially the gram sabha, as envisaged in Article 40.

The Constitution gives powers to the High Courts and the Supreme Court to issue writs to strike down unconstitutional legislative and executive action.[40] According to the courts, "[t]he right to life includes the right to live with human dignity and all that goes with it, namely, the bare necessaries of life such as adequate nutrition, clothing and shelter and facilities for reading, writing and expressing oneself in diverse forms, freely moving about and mixing and co-mingling with fellow human beings. The magnitude and components of this right would depend upon the extent of economic development of the country, but it must, in any view of the matter, include the bare necessities of life and also the right to carry on such functions and activities as constitute the bare minimum expression of the human self".[41] They have also said that "water is the basic need for the survival of human beings and is part of the right to life and human rights as enshrined in Article 21 of the Constitution of India and can be served only by providing source of water where there is none".[42]

The court in the context of water pollution mandated the cleaning up of water sources, including rivers,[43] the coastline,[44] and even tanks and wells. The court said: "It is important to notice that the material resources of the community like forests, tanks, ponds, hillocks, mountains, etc. are nature's bounty. They need to be protected for a proper and healthy environment, which enables people to enjoy a quality of life which is the essence of the guaranteed right under Article 21 of the Constitution of India".[45] The unregulated discharge of effluents has led the court to issue mandatory directions for cleaning up by the polluter and restitution of the soil and groundwater.[46] The court has also applied the "precautionary principle" to prevent the potential pollution of drinking water sources resulting from the setting up of industries in their vicinity.[47]

The court, recognizing that water is a community source to be held by the state in public trust, and in recognition of its duty to respect the principle of inter-generational equity declared that "our legal system – based on English common law – includes the public trust doctrine as part of its jurisprudence. The state is the trustee of all natural resources which are by nature meant for public use and enjoyment. The public at large is the beneficiary of the seashore, running waters, air, forests and ecologically fragile lands. The state as a trustee is under a legal duty to protect the natural resources. These resources meant for public use cannot be converted into private ownership".[48]

There are numerous obstacles to be surmounted before effective and equal access to water becomes a certainty. The disputes between states, sometimes even erupting into social conflicts,[49] are referred to the Inter-State Water Disputes Tribunal constituted under Article 262, takes years to reach an award[50] that, in the end, are often not honored by the parties.[51] With the states prioritizing political convenience over their obligations to honor their commitments to respect the rule of law, the rights of farmers to river water for irrigation are often violated.

At the community level, conflict exists with the state and village level administrative bodies, over the control of water sources, rendering the constitutional scheme of decentralization of responsibilities over water sources unworkable. The state can also legislatively avoid any contractual obligation to assure availability of fixed quantities of water for irrigation purposes to cultivators,[52] for purposes of providing "benefit to a larger area and more people".[53] On provision of equal access to water at the community level, caste-based inequality can prevent socially disadvantaged groups from accessing common water resources, or render water unusable, which is punishable in law.[54]

The traditional village water governance system is marginalized or destroyed, with the state legislations failing to recognize them. Changing priorities of the state generates a livelihood crisis for some users of water, when communities are denied traditional access rivers and reservoirs.[55] The claims by industrial units that their right to carry on trade and business be accorded priority over the right to a clean environment, resulting in conflicts have frequently been resolved by the court favoring the latter. The court, drawing extensively on international environmental law employs the "polluter pays principle"[56] and the "precautionary principle".[57] The court has asserted that industry "has no right to destroy the ecology, degrade the environment and pose a health hazard. It cannot be permitted to expand or even continue with the present production, unless it tackles by itself the problem of pollution created by it".[58] "The precautionary principle" replaced the "assimilative capacity"[59] of the environment where the "burden of proof, in environmental cases, as to the absence of injurious effects of the actions proposed, is placed on those who want to change the status quo" and "an order of exemption carelessly passed, ignoring the precautionary principle, could be catastrophic".[60] However, the Supreme Court, in some cases,[61] did not take recourse to the precautionary principle, stating that "once the state government has taken all precautions to ensure that the impact on the environment is transient and minimal, a court will not substitute its own assessment in place of the opinion of persons who are specialists and who may have decided the question with objectivity and ability".

There is a major thrust to corporatize the distribution of water for drinking purposes, in both the urban and rural areas.[62] The bottled-water industry is a high-growth sector, adding to the already stressed availability of water to communities for their basic survival needs. There was also an instance of sale of even a river.[63] In instances of industries using water as the major raw material, the extent to which panchayats can control the activities of corporations has come up, with the state questioning the legality of the domain of panchayats.[64]

(a) There is a distinct trend with the replacement of the state's obligation to ensure availability to the community of basic minimum quantities of affordable quality water, with water as a commodity of commerce. Strangely, water is also seen as a coercive instrument in issues like curbing population growth[65] and forcible displacement.[66] This is a violation of equal access to common resources. The courts have frequently attributed disproportionate blame on poor households for pollution of water bodies, often forcibly evicting them in favor of industrial units.[67] Utilitarian logic often pervades government decisions, rather than issues of equity, justice, ecology and sustainability. The larger good is often cited by the courts to explain away large scale displacement, with rehabilitation of the oustees itself being interpreted as better in terms of amenities available to them and, in the case of tribals, their "gradual assimilation in the mainstream of society will lead to betterment and progress." The courts have sometimes defended the state in such instances, by saying, for instance, that 'when a decision is taken by the government after due consideration and full application of mind, the court is not to sit in appeal over such decision' and explains such displacement as necessary for the "larger good".[68]

The Plachimada struggle therefore is consistent with and is an outcome of, larger governance issues pertaining to water, as is evident from the legislative and judicial history of water governance in this country.

Towards People's Governance of Water

In its Report of the Expert Group on Groundwater Management and Ownership,[69] the Planning Commission concluded in Section 8 that:

(a) In limiting extraction, the first thing that needs to be addressed is the legal regime. Under the current legal regime, which dates back several decades, the landowner is given the right to capture an unlimited amount of groundwater from beneath his/her land, without being liable for injury to adjacent landowners caused by excessive or harmful pumping. By relying on this regime, our historical approach has been to exercise little control on groundwater pumping. The approach may have been adequate when the overall demand for groundwater in any given aquifer was limited, but with the threat of large-scale withdrawals looming large, there is a need for change in approach.

(*b*) The recent court rulings have emphasized the role of state as the trustee of all natural resources, including groundwater, which by nature are meant for use and enjoyment of public at large. Pursuing this position, courts have held that the state has got a duty to protect groundwater against excessive exploitation, and the inaction of the state in this regard will tantamount to infringement of the right to life of the people guaranteed under Article 21 of the Constitution of India.

(*c*) The attempt must therefore be to balance the landowner's right to capture groundwater, with the public interest in managing groundwater resources for all users, including the environment, and to ensure that both the present and future needs of the communities dependent upon these resources are accounted for.

(*d*) In the light of the above, the Group emphasizes the need for all states to introduce a modified groundwater legislation, encompassing inter alia, the role and responsibility of water user groups and the government. The involvement of Panchayati Raj institutions should be a key part of the strategy.

(*e*) Further, given the enforcement problems relating to prohibitive measures, greater reliance needs to be placed on community management of the resource, supported by adequate technical inputs, complementary institutional changes and appropriate incentives (such as a subsidy regime for micro-irrigation), rather than on controls by state, for reasons stated earlier. State legislations can, of course, strengthen such strategy by endorsing community action, supportive institutions and use of technical inputs and incentives.

The law is yet to make clarifications on matters related to rights over water and governance of water. The sources of law are the

Constitution of India, statutory provisions, case laws or court precedents, doctrines and principles deriving from the British common law system, international agreements, personal law, and customary law and practices. The right to groundwater is linked to right to land, supported by the legal construction under the Indian Easements Act of 1882, by which there is almost absolute ownership of groundwater by the property holders. At the same time, the state as an "eminent domain" has power over natural resources – the British colonizers introduced the concept of "eminent domain" of the state, where all resources that were "unclaimed" or "not used," belonged to the British Empire. These laws still remain as statutory laws, constituting the framework for governance, creating both a barrier for the democratization of governance of resources and a wedge between the state and the people, leading to increasing conflicts.

In the federal set up of the Constitution, such subject matters as "water" and land are within the legislative power of the states under Article 246 and List II of the Seventh Schedule. The subject of water is expressed in terms of "water supplies, irrigation and canals, drainage and embankments, water storage and water power" under Entry 17 of List II. Apart from a reference to an exception for inter-state rivers, it is the state that is the exclusive authority to make laws with respect to water. The Ministry of Water Resources had issued model bills for the states to enact, centralizing the authority with the state governments. Under this, State Groundwater Authorities are to manage and develop groundwater resources.

On the other hand, the 73rd constitutional amendment Act of 1992 on Panchayats provides for "drinking water" and "minor irrigation, water management and watershed development" in the 11th Schedule as subject matters over which responsibility can be devolved from the state to the village level. Subsequently, it is up to each state to pass regulations on the authority of the panchayats on the 29 listed subjects. Accordingly, the duties of the village panchayats include, "maintenance of traditional drinking water sources, preservation of ponds and other water tanks, and setting up and management of water supply schemes".[70]

Matters such as natural resource management, including water, are very much a community domain exercised through customary laws and practices. The Constitution of India recognizes customs and customary practices. The term "law" in Article 13 includes "customs" and "usages" having the force of law, but not infringing any of the fundamental rights conferred by Part III of the Constitution. In reality, customary laws are subjected to being in consonance with the statute-made laws, which are still largely a colonial legacy. The state under Article 39(b) is duty bound to sub-serve the common good. Part IV-A of the Constitution imposes a duty (and not authority) on the citizens to protect and improve the natural environment including forests, lakes and rivers which are explicit directives to be promoted through legislations.

It is in the Fifth Schedule Area that the Panchayat Raj (Extension to Scheduled Areas) Act, 1996 made a decisive departure, introducing a framework to dismantle the colonial legislative frame with comprehensive powers for "traditional management practices of community resources," "to safeguard and preserve community resources" and "planning and management of minor water bodies." "Community resources" though not defined in the Act naturally include water. The assertion of the constitutional goal of the community or a people-centric state, rather than the state versus people paradigm, has resulted from the adivasis' struggles. Consequently, the gram sabhas, rather than the panchayat or the state, is to be in primary command, with restrictions on the higher authorities not to assume the powers and authority of any lower panchayat or gram sabha, legally paving the way for a people's participatory democratic governance system.

Yet, the dichotomy between the sovereign rights of the people and the state to natural resources persists, a contradiction that, in many places, has led the state to wage war on its own people. The sovereign right of the nation, that is, of her people, is sometimes clearly on a collision course with the state. The statutory laws arising from the constitutional framework and those that are a colonial legacy are on a collision course. There is a strong reluctance to dismantle the colonial frame, with a nexus between the bureaucratic structure and the dominant political and economic class. But this is possible with sustained mass struggles, and the movement has once again demonstrated success in the case of forest resources, resulting in the enactment of the "The Scheduled Tribes and Other Traditional Forest Dwellers (Recognition of Forest Rights) Act, 2006" where the rights, both individual and collective, are to be determined by the gram sabha, who is also empowered to govern its "Community Forest Resource".[71]

Democratization requires a decisive move away from *decentralization* to a *non-centralized* people's participatory governance system. The gram sabha, the fundamental and primary natural unit of governance in the panchayat structure, shall be the competent authority to determine the use, control and management of groundwater and minor water bodies within its geographical jurisdiction. The gram sabha shall have command over groundwater for traditional management practices of groundwater resources, safeguarding and preserving groundwater resources, planning and management of minor water bodies and management of water supply schemes. The gram sabha, while exercising its powers to safeguard and preserve groundwater resources and minor water bodies, shall have the power to ensure safe and adequate water to its members. While exercising this authority, the gram sabha shall reasonably restrict the use of groundwater and minor water bodies to ensure safety and adequacy of water and reasonably restrict or ban their use if the gram sabha so considers that such use constitutes a threat to quantity and/or quality of water. The gram sabha shall have the authority to take necessary steps and penalize violators in the case of violations that threaten the quality and quantity of water resources. All development activities falling within the geographical jurisdiction of the gram sabha shall be carried out with the express consent and approval of the gram sabha, with a view to protect, conserve, sustainably use and develop the said water resources. A resolution by the gram sabha to cancel such consent shall automatically lead to cancellation of licence/ approval issued by the appropriate bodies/authorities.

To enable the gram sabha to carry out these responsibilities and functions to the fullest extent, the gram sabha shall call forth the services and assistance of the State Groundwater Authority, the Groundwater Department, the State Pollution Control Board, the village panchayat, the block panchayat and/or the district panchayat, as the case may be, and other relevant bodies, which shall be made available.

To enable the gram sabha to exercise this natural right, it is therefore necessary that relevant state laws such as (a) the state Panchayat Raj Acts; (b) the state groundwater Acts; (c) the Water Pollution Prevention and Control Acts; and (d) any other relevant laws as are applicable, are to be amended. The 73rd Amendment, should be reviewed and amended, conferring the command over natural resources to the gram sabhas. A decisive shift from the predominantly protest mode, to an assertion of the people's command over resources, is the turning tide that can usher in much needed, sweeping changes in water governance.

Significantly the Planning Commission is involved in drafting a "model bill for the protection, conservation, management and regulation of groundwater" in 2011. This is part of the exercise towards developing the Twelfth Five-Year Plan (2012–17). But the Ministry of Panchayati Raj went much beyond mere tinkering with laws by proposing amendments to Article 243 of the Constitution for strengthening the local bodies[72] in March 2010. This, among other things, proposes the creation of 13th Schedule, listing the subject matters exclusively under the "powers and functions of gram and ward Sabha" on similar lines as the governance powers in Fifth Schedule (see earlier). "Competence to safeguard and preserve the community resources including land, water, forest and minerals" besides "planning and management of minor water bodies" are listed in the Schedule. The Ministry also proposed strengthening of the self-governance powers of the gram sabha in Fifth Schedule areas [73] through amendments and introducing similar structures and powers in the Sixth Schedule areas in the north-eastern states through constitutional amendments.[74] These, if taken to its logical end, would mean a radical overhaul of the governance structure in the country, a definite move away from the colonial state-centric to finally a democratic people-centric structure. The battle for control over natural resources, it seems, has reached a watershed under the intense heat of the spreading conflicts.

Notes

1. Case Law refers to the decisions and interpretations made by judges while deciding on the legal issues before them which are considered as the common law or as an aid for interpretation of a law in subsequent cases with similar conditions. Case laws are used by advocates to support their views to favor their clients and it also influences the decision of the judges. See http://www.legal-explanations.com/definitions/case-law.htm (accessed August 1, 2011).
2. The text of the Bill is available at http://www.niyamasabha.org/bills/12kla/plachimada%20victims.pdf (accessed August 4, 2011).
3. "Kerala passes Bill, asks Coca-Cola to pay INR 216 crore to Plachimada victims", *The Financial Express*, 25 February 2011, http://www.financialexpress.com/news/kerala-passes-bill-asks-cocacola-to-pay-r216-crore-to-plachimada-victims/754506/ (accessed August 4, 2011).
4. The Bill was sent to the President by the Governor of Kerala on March 30, 2011 through the Ministry of Home Affairs. The Ministry forwarded the Bill to the Ministry of Rural development, Ministry of Agriculture, Ministry of Law and Justice, Ministry of Water Resources, Ministry of Food Processing Industries, and Ministry of Environment and Forests for comments. None of these Ministries raised any objections and supported the Bill with some Ministries suggesting provisions to strengthen the Bill. However, contrary to cabinet guidelines, the Ministry of Home Affairs, instead of forwarding the Bill to the President for assent, sent the Bill to Coca Cola on July 12, 2011 for their comments which was forwarded to the Governor of Kerala on July 8, 2011. The Kerala government rebutted all the points raised by Fali S Nariman, Coca Cola's legal advisor, and sent it to the Governor for forward to the Ministry of Home Affairs on November 17, 2011. The Members of Parliament and the Chief Minister too had appealed to the President to expedite the matter. Presidential assent is still awaited as on October 2013.
5. Choudhury, Uttara. *Coke not to sell Kerala plant*, June 18, 2007. Available online at http://www.dnaindia.com/report.asp?NewsID=1103967 (accessed November 13, 2007).
6. The "Report of The High Power Committee to Assess the Extent of Damages Caused by The Coca-Cola at Plachimada Plant and Claiming Compensation" is available at http://www.groundwater.kerala.gov.in/english/index.htm (accessed August 4, 2011).
7. "Coca Cola must pay damages, says panel", *The Hindu*, March 15, 2010. Available online at http://www.hindu.com/2010/03/15/stories/2010031561360700.htm (accessed August 30, 2010).
8. Water Resources (GW) Department, G.O (MS) No. 37/2010/WRD dated July 3, 2010, Water Resources (Ground Water) Department, Government

of Kerala, available at http://www.groundwater.kerala.gov.in/english/pdf/gwd_37_2010.pdf (accessed on August 4, 2011).
9. *Democratic control of groundwater sought.* The Hindu, October 13, 2008. http://www.hinduonnet.com/thehindu/thscrip/print.pl?file=2008101352100500.htm&date=2008/10/13/&prd=th& (accessed November 19, 2008).
10. For the Terms of Reference, see G.O(Rt) No.425/2009/WRD dated March 17, 2009, Water Resources (Ground Water) Department, Government of Kerala, available at http://www.groundwater.kerala.gov.in/english/pdf/go_tor.pdf (accessed August 4, 2011).
11. Located to the east of Palakkad district, the panchayat has an area of 60.91 sq kms in the Chittoorpuzha sub-basin of the Bharatapuzha river basin. It borders Chittoorpuzha river in the north, the Vandithavalam-Meenakshipuram state highway and Pattancherry Panchayat in the south, Chittoor-Thathamangalam in the west, while the eastern boundary is parallel to the meandering course of the Chittoorpuzha river, bordering the Coimbatore district of Tamil Nadu state.
12. Chakkiliyan, Parayan, Pallan and Cheruman are some of the important Dalit communities found here.
13. 100 cents = 1 acre = 43,560 ft^2.
14. *Resource Map Report 2001* as quoted in Nair, K.N., Antonyto Paul and Vineetha Menon. *Water Insecurity, Institutions and Livelihood Dynamics: A Study in Plachimada, Kerala, India*, Daanish Books, Delhi and Centre for Development Studies, Thiruvananthapuram, 2008, pp.57.
15. The Parambikulam-Aliyar Project (PAP),signed in 1970 (with retrospective effect from 1958) between Tamil Nadu and Kerala was commissioned in 1974. The agreement was to share the waters of the Bharathapuzha, Chalakkudypuzha and Periyar rivers, which originate in the Anamudi Hills of Tamil Nadu, and flow through Kerala into the Arabian Sea. The water is collected in a series of nine reservoirs for irrigation.
16. *Estimation of Groundwater: Draft for Domestic, Agriculture and Industrial Use in Plachimada area, Palakkad District, Kerala*, Central Ground Water Board, Ministry of Water Resources, Kerala Region, Trivandrum, May 2005, pp. 13 & 19.
17. Coca-Cola, registered as a trademark in 1887, operates in over 200 countries worldwide, generated more than 70 percent of its income outside the US by 2003. Coca-Cola, headquartered in Atlanta, USA, re-entered India in 1993 after being expelled in 1977 for violating the Foreign Exchange Regulation Act (FERA), under which it was to reduce its equity stake. HCCB, registered in 1993 as a subsidiary of Coca-Cola, invested more than US$1 billion between 1993 and 2003, establishing 27 wholly-owned bottling operations, supplemented by 17 franchisee-owned bottling operations and a network of 29 contract-packers (who package and label the products of other company) to manufacture a range of products.

18. It has the Meenkara dam reservoir about 3 km to the north, and the Kambalathara and Vengalakkayam storage reservoirs a few hundred metres to the west. Barely 10 metres north of the factory compound, is the Moolanthodu main canal from the Moolathara barrage. The main Chittoorpuzha river itself flows 2 kms due north of the plant.
19. Under Section 232, 234 and 254 of the Kerala Panchayat Raj Act 1994.
20. Central Ground Water Board, op cit, p. 19.
21. For a detailed narrative on the Plachimada Struggle, refer Bijoy, C.R. *Kerala's Plachimada Struggle. A Narrative on Water and Governance Rights*, Economic and Political Weekly, October 14, 2006, pp. 4332-39. Also available at http://www.epw.org.in/epw/uploads/articles/9348.pdf (accessed November 13, 2007).
22. 'On coming to know from the media about the BBC report on high concentration of cadmium and lead in the sludge from the industry and in nearby wells, the officers of District Office of the Board at Palakkad collected random samples of sludge and effluent and those were analysed in the Board's Laboratory at Ernakulam. The result indicated a cadmium concentration of 201.8 mg/kg in the sludge'. *Presence of Heavy Metals in Sludge Generated in the Factory of M/S Hindustan Coca-Cola Beverages Pvt. Ltd, Palakkad, A Study Report*, Kerala State Pollution Control Board, September, 2003.
23. Constituted to monitor the implementation of the Supreme Court's order of October 14, 2003.
24. Anand, S. *Don't Poison My Well*, Outlook India, May 16, 2005, http://www.outlookindia.com/article.aspx?227376 (accessed August 4, 2011).
25. *Ground Water Resources in Plachimada Coca Cola stores toxics for future generations*, Hazards Centre, New Delhi and People's Science Institute, Dehradun, June 2006.
26. O.P. No. 13513/2003 praying for calling for records, leading to the passing of the resolution.
27. W.P.(C) No. 34292 of 2003.
28. W.A. No. 2125 of 2003 dated 17 December 2003 challenging the judgment of the single judge.
29. KSPCB order PCB/PLKD/CE/32/99 dated 19.08.2005.
30. The team comprised the Centre for Water Resources Development and management, Kozhikode as the Convenor; Kerala State Ground Water Department; Kerala State Pollution Control Board and HCCB.
31. The non-renewal of contracts with Coca-Cola by the universities prompted Coca-Cola to get The Energy and Resource Institute (TERI) to assess its operating plants in India. The Plachimada plant was not one of the plants studied as it was not in operation at the time. For a commentary on this study, refer Bijoy, C.R. *Good Water, Bad Cola?*, Tehelka *Magazine*,

Vol 5, Issue 4, Feb 2, 2008. Also available at http://www.tehelka.com/story_main37.asp?filename=cr020208Good_Water.asp (accessed August 30, 2010).
32. For instance, they launched an agitation to get their due share from Tamil Nadu, as per the Parambikulam Aliyar Project agreement, resulting in an inter-state dialogue at the highest level, and deepened existing wells.
33. The struggle culminated with the government formally agreeing to provide land to landless Adivasis and to include Adivasi habitations in the Scheduled Area under the V Schedule of Article 244 of the Constitution. For more details, see: Bijoy, C.R and K. Raviraman. *Muthanga: The Real Story. Adivasi Movement to Recover Land, Economic and Political Weekly,* May 17, 2003, pp.1975-82. Available at http://www.epw.org.in/epw/uploads/articles/2819.pdf (accessed November 13, 2007).
34. The Coastal Regulation Zone (CRZ) Notifications. Available at http://envfor.nic.in/legis/crz/czma.html.
35. Such as The Northern India Canal and Draining Act, 1873, The Bombay Irrigation Act, 1879, The Rajasthan Irrigation and Drainage Act, 1954, Karnataka Irrigation Act, 1965.
36. Such as The Kumaon and Garhwal Water (Collection, Retention and Distribution) Act, 1975.
37. Such as The Uttar Pradesh Water Supply and Sewerage Act, 1975.
38. Such as Madras Metropolitan Area Ground Water (Regulation) Act, 1987, Kerala Ground Water (Control and Regulation) Act, 2002, Karnataka Ground Water (Protection and Regulation for Drinking Water) Act, 2003.
39. The Water (Prevention and Control of Pollution) Cess Act, 1977.
40. Under Article 13 of the Constitution, any law that is inconsistent or contrary to the fundamental rights enumerated in Part III of the Constitution would be liable to be declared as void. The Supreme Court of India, under Article 32 of the Constitution, and each of the High Courts, under Article 226 of the Constitution, can issue writs to declare the unconstitutionality of statutes that are challenged by way of Writ Petitions.
41. Francis Coralie Mullin v. The Administrator, Union Territory of Delhi 1981 (2) SCR 516.
42. Narmada Bachao Andolan v. Union of India (2000) 10 SCC 664 at 767.
43. For orders relating to the pollution on the river Ganga, see M.C. Mehta v. Union of India AIR 1988 SC 1037, 1115 and (1997) 2 SCC 411. For an important decision regarding closure of a hotel resort polluting the Beas river in Himachal Pradesh, see M.C. Mehta v. Kamal Nath (1997) 1 SCC 388.
44. S. Jagannath v. Union of India (1997) 2 SCC 87.
45. In Hinch Lal Tiwari v. Kamala Devi (2001) 6 SCC 496.
46. Bhavani River-Shakti Sugars Ltd. (1998) 6 SCC 335. In Indian Council for Enviro-Legal Action v. Union of India (1995) 3 SCC 77, a compensation

package was worked out for farmers when their only source of irrigation, a river in Andhra Pradesh, was polluted by discharge of untreated effluents by industries alongside its banks.
47. A.P Pollution Control Board v. Prof. M.V Nayudu (1999) 2 SCC 718 and A.P Pollution Control Board (II) v. Prof. M.V Nayudu (2001) 2 SCC 62.
48. M.C. Mehta v. Kamal Nath (1997) 1 SCC 388.
49. The disputes between Tamil Nadu and Karnataka over the Cauvery river led to riots and loss of life and property in both the states, were referred to a similar tribunal in 1990, which gave a final verdict only in 2007. For orders of the Supreme Court directing the setting up of tribunals in both the states determining of claims of compensation by the victims of the riots, see Ranganathan v. Union of India (1999) 6 SCC 26.
50. The dispute between the States of Gujarat, Madhya Pradesh, Maharashtra and Rajasthan over the river Narmada referred to the Inter-State Water Disputes Tribunal constituted under Article 262, leading to an award a decade later, only led to further disputes. A final award was made in 1977. Refer Narmada Bachao Andolan v. Union of India (2000) 10 SCC 664 at 767.
51. Disputes between Punjab, Haryana and Himachal Pradesh over the Ravi and Beas rivers proved problematic, with Punjab refusing to honor the tribunal award. Haryana approached the Supreme Court for its enforcement. State of Haryana v. State of Punjab (2004) 12 SCC 673.
52. The Parambikulam Aliyar Project of the State of Tamil Nadu was to ensure supply of water for agricultural operations in some areas of Coimbatore district. Although the state government entered into an agreement with the cultivators to assure supply of a minimum quantity of water for irrigation purposes, it passed a legislation 10 years later, seeking to expand the area over which the said water would be made available for irrigation purposes.
53. Parambikulam A.P.O.A. Association v. State of Tamil Nadu (1997) 7 SCC 626 at 632.
54. Sec.3 (xiii) of the Scheduled Castes and Scheduled Tribes (Prevention of Atrocities) Act, 1989 which makes a punishable offence the act of a person, not being a member of a Scheduled Caste or Scheduled Tribe, who "corrupts or fouls the water of any spring, reservoir or any other source ordinarily used by the members of the Scheduled Castes or Scheduled Tribes so as to render it less fit for the purpose for which it is ordinarily used." A similar provision is Sec. 4 (iv) of the Protection of Civil Rights Act, 1955 which renders punishable the act of a person who, on the ground of "untouchability," enforces against any other person any disability with regard to "the use of, or access to, any river, stream, spring, well, tank, cistern, water-tap or other watering place, or any bearing ghat."

55. As for instance, denial of fishing rights imposed under the Wildlife (Protection) Act, 1972 in a protected area.
56. Indian Council for Enviro-Legal Action v. Union of India (1996) 3 SCC 212 in which residents of Bichhri village in Rajasthan who were denied access to drinking water on account of pollution of the wells by seepage of toxic untreated wastes produced by chemical factories, were held entitled to restitution and compensation. Also S. Jagannath v. Union of India (1997) 2 SCC 87, In Re: Bhavani River-Shakti Sugars Ltd. (1998) 6 SCC 335 and in Indian Council for Enviro-Legal Action v. Union of India (1995) 3 SCC 77 the Supreme Court applied the precautionary principle to strike down the notification issued by the Government of Andhra Pradesh exempting an oil industry located in the vicinity of two major water reservoirs.
57. In the Vellore Citizens Welfare Forum vs. Union of India (1996) 5 SCC 647 the Supreme Court held that the precautionary principle is part of the law of the land, which means that where there are threats of serious and irreversible damage, lack of scientific certainty should not be used as a reason for postponing measure to prevent environmental degradation.
58. Vellore Citizens Welfare Forum v. Union of India, (1996) 5 SCC 647.
59. The capacity of a natural body of water to receive wastewaters or toxic materials without deleterious effects and without damage to aquatic life or humans who consume the water.
60. A.P. Pollution Control Board (II) v. Prof. M.V. Nayudu (2001) 2 SCC 62 at 79. More recently, the Supreme Court declared as illegal the action of the Government of Karnataka excluding the land brought under mining by the Kudremukh Iron Ore Company Ltd. excluded from the purview of notification issued under s.35 (1) of the Wildlife (Protection) Act 1972 declaring the Kudremukh National Park to be a national park: T.N. Godavarman Tirumalpad v. Union of India (2002) 8 SCALE 204. In MC Mehta vs. Union of India (1997) 2 SCC 353 the Supreme Court held that the 'onus of proof' is on the developer/ industrialist to show that his action is environmentally benign.
61. Essar Oil Ltd. v. Halar Utkarshi Samiti (2004) 2 SCC 392.
62. While Maharashtra has enacted a law which paves the way for privatization, the draft legislation in Delhi has met with stiff resistance from water users.
63. The exclusive right to 23.5km of the Sheonath River was handed over to a private company in Chhattisgarh under the build-own-operate-transfer agreement by a state owned industrial development corporation. The company was to sell the waters to the state owned corporation which in turn would sell it to the ultimate users in the nearby industrial township. This impacted the fisher folk who were barred from fishing in the river and the small farmers who depended on the river. Following protests, the government announced the cancellation of the agreement.

64. Judgment related to the Hindustan Coca-Cola Beverages (P) Ltd. v. Perumatty Grama Panchayat dated December 16, 2003 of K. Balakrishnan Nair, J. of the Kerala High Court in Writ Petition (C) No. 34292 of 2003 and Judgment dated April 7, 2005 of the Division Bench of the Kerala High Court in Writ Appeal No. 215 of 2004. The case is in the Supreme Court.
65. Maharashtra Water Resources Regulatory Act 2005 in Sec.12 (11) states: 'Notwithstanding anything contained in this Act, a person having more than two children shall be required to pay one and half times the normal rates of water charges fixed' which places an unconscionable restriction on the availability of affordable water by linking the payment for water to the size of the family.
66. In cases of land acquisitions and slum demolitions.
67. In the order dated March 3, 2003 of the Delhi High Court in Writ Petition 2112 of 2002 (Wazirpur Bartan Nirmata Sangh), the Court ordered the removal of unauthorized encroachments along the banks of the river Yamuna flowing through Delhi, considering them as the main cause of the pollution, while the city's industrial units, many of them unauthorized, the real cause of pollution by discharging of large quantities of untreated effluents into it, were left untouched.
68. Narmada Bachao Andolan v. Union of India (2000) 10 SCC 664.
69. See 3, 5, 6, 10 and 11 of Section 8 'Summary and Conclusion' of the '*Report of the Expert Group on Ground Water Management and Ownership*' September 2007, Planning Commission, New Delhi 110001 http://planningcommission.nic.in/reports/genrep/rep_grndwat.pdf (accessed August 6, 2011).
70. As for instance the Third Schedule under Sec. 166 of the Kerala Panchayat Raj Act 1994.
71. For an overview see Bijoy, C.R. *Forest Rights Struggle. The Adivasis Now Await a Settlement,* American Behavioral Scientist, Sage Publications, Vol. 51, No. 12, August 2008, pp. 1755–73. 10.1177/0002764208318929 http://abs.sagepub.com (accessed November 19, 2008).
72. Available at http://panchayat.nic.in/viewContentItem.do?method=viewItem&itemid=5069&ptltid=3947&folderid=3956 (accessed August 5, 2011).
73. Through amendment to the Panchayats (Extension to Scheduled Areas) Act, 1996 available at http://panchayat.nic.in/viewContentItem.do?method=viewItem&itemid=5070&ptltid=3947&folderid=3956 (accessed August 5, 2011).
74. Amendment in the Sixth Schedule, Part IX (Article 243A, 243M) Article 339, and Article 275(1), available at http://panchayat.nic.in/viewContentItem.do?method=viewItem&page=sr&itemid=5269 (accessed August 5, 2011).

References

Bijoy, C. R. 2008. "Good Water, Bad Cola?," *Tehelka*, 5(4).
———. 2006. "Kerala's Plachimada Struggle: A Narrative on Water and Governance Rights,"*Economic and Political Weekly*, 41(41): 4332–39.
Hazards Centre and People's Science Insitute. 2006. *Ground Water Resources in Plachimada Coca Cola Stores Toxics for Future Generations.* New Delhi and Dehradun: Hazards Centre and People's Science Insitute.
Investigations on The Extractions of Groundwater By M/s Hindustan Coca-Cola Beverages Private Limited at Plachimada, Final Report, The Investigation Team, Constituted vide Order WA/2125/2003, dated 19.12.2003 by the Honourable High Court of Kerala, February 11, 2005.
Muralidhar, S. 2006. "The Right to Water. An Overview of the Indian Legal Regime,"in Eibe Riedel and Peter Rothen (eds), *The Human Right to Water.* Berlin: Berliner Wissenschafts-Verlag,.
Nair, K. N., Antonyto Paul and Vineetha Menon. 2008.*Water Insecurity, Institutions and Livelihood Dynamics: A Study in Plachimada, Kerala, India.* Thiruvananthapuram: Daanish Books in collaboration with Centre for Development Studies.
Kerala State Pollution Control Board. 2003. *Presence of Heavy Metals in Sludge Generated in the Factory of M/s Hindustan Coca-Cola Beverages Pvt. Ltd, Palakkad, A Study Report.*
Raman, K. Ravi. 2005. "Corporate Violence, Legal Nuances and Political Ecology. Cola War in Plachimada,"*Economic and Political Weekly*, 40(25): 2481–86.
Planning Commission. 2007. *Report of the Expert Group on Ground Water Management and Ownership.* New Delhi: Planning Commission.
Government of Kerala. 2010. *Report of The High Power Committee to Assess the Extent of Damages Caused by The Coca-Cola at Plachimada Plant and Claiming Compensation.*
Thapliyal, B. K., S. S. P. Sharma and Hemanth Kumar (eds). 2008. *Democratisation of Water.* New Delhi: Serials Publishers for National Institute of Rural Development (NIRD).

About the Editors

N. C. Narayanan is Associate Professor in CTARA, Indian Institute of Technology Bombay (IITB). He has a PhD in Development Studies from the Institute of Social Studies, The Hague, The Netherlands and has taught at the Institute of Rural Management, Anand and worked as the Executive Director of SaciWATERs, Hyderabad. He is the author of *Against the Grain: The Political Ecology of Land Use in a Kerala Region, India* (2003), co-author of *TINA and the Milk: Southern Perspectives on Sustainability in the Netherlands* (2002) and editor of *Where to go fromhere? State, Natural Resource Conflicts and Challenges to Governance* (2008). His areas of research interest are water policy, governance, conflicts and the political economy of environmental degradation.

S. Parasuraman is Director, Tata Institute of Social Sciences (TISS), Mumbai, India. He has over 25 years of experience as a teacher, trainer, activist, administrator, and development worker. He obtained his PhD in Demography from the University of Mumbai, Mumbai. He has held key positions in international organizations such as Asia Regional Policy Coordinator, ActionAid Asia; Senior Advisor to the Commission, and Team Leader of the Secretariat; World Commission on Dams; and Programme Director, Oxfam GB, India Programme.

Rajindra Ariyabandu is development researcher with a special interest in water resources management and community development. He has experience of 20 years of working in various capacities with government and non-governmental organizations and international research institutions. He has been actively involved in water policy development and coordination of regional water resources management for the last 10 years. He has extensive field research experience in the implementation of low-cost, community-managed water utilities for the rural poor.

Notes on Contributors

C. R. Bijoy is an activist and independent researcher involved in issues of governance and natural resources. He is presently engaged with the Campaign for Survival and Dignity (a coalition of mass organizations of forest dwellers on forest rights) and a few other local resource related struggles.

Vasudha Chhotray is Senior Lecturer, School of International Development, University of East Anglia, United Kingdom. A political scientist with varied interdisciplinary research interests, she works mainly on governance with a focus both on state and civil society, and the macro–micro links between institutions, discourses and policies. More recently, she has focused on studying disasters, in particular, the long-term trajectories of vulnerability and recovery in their aftermath. Through this research, she has also become actively interested in social justice theory and the state's engagement with social justice with respect to disasters.

Mahfuzul H. Chowdhury is Director and Professor of Political Science, Social Science Research Institute, University of Chittagong, Bangladesh. His research interests include democratization, governance, US politics/foreign policy and South Asian politics. His noted publications include *Democratisation in South Asia: Lessons from American Institutions* (2003) and the article "Violence Politics and the State in Bangladesh," *Journal of Conflict Security & Development.*

Hamidul Huq is Chairman, Institute of Livelihoods Studies (ILS). A sociologist by training, he obtained his PhD in Rural Development Sociology from Wageningen University in 2000. He has strong knowledge in interdisciplinary education and research. He was actively involved in the leadership level of FAP Protest Campaign during the 1990s and has been devoted to research and policy advocacy in the water sector. Currently, he is engaged in research and advocacy in the fields of water, ecosystem services, climate change and livelihoods in the coastal zone.

S. Mohammed Irshad is Assistant Professor, Jamsetji Tata Centre for Disaster Management, Tata Institute of Social Science (TISS), Mumbai. He obtained his PhD from the University of Kerala, specializing in water governance.

K. J. Joy is Senior Fellow, Society for Promoting Participative Ecosystem Management (SOPPECOM), Pune. He has been an activist–researcher for more than 20 years and has a special interest in people's institutions for natural resource management, both at the grassroots and policy levels. His other areas of interest include drought and drought proofing, participatory irrigation management, river basin management and multi-stakeholder processes, watershed-based development, and water conflicts and people's movements.

Seema Kulkarni is Senior Fellow, SOPPECOM, Pune. Her key area of interest within SOPPECOM, is gender and natural resource management. She has over 10 years of experience of working in drought prone rural areas of Maharashtra and is also closely associated with the women's movement in Maharashtra. She has worked with the Stree Mukti Sangharsh Chalwal (women's liberation struggle) in western Maharashtra, specifically on the question of single and deserted women. She has authored the booklet *Intensive Cultivation on Small Plots* and co-authored *Women, Water and Livelihoods* (2007).

Suhas Paranjape is Secretary and Senior Fellow, SOPPECOM, Pune. He has participated as a core team member and consultant in many action research studies and pilot projects in the areas of participatory management of natural resources, especially in the field of participatory irrigation management, development theory, economic theory, water sector, and alternative development models. He has also actively participated in different movements, such as the People's Science Movement, and has worked full time for Shramik Sanghatana, the Adivasi agricultural labourers' movement in Nandurbar, Maharashtra. He has written extensively and co-authored a number of books on these themes.

Ruana Rajepakse is a lawyer based in Sri Lanka whose areas of practice include civil law, public law and environmental law. She is the author of *An Introduction to Law in Sri Lanka* and *A Guide to Current Constitutional Issues in Sri Lanka*. She was also Editor, *South Asian Environmental Law Reporter* (1994–2000) and has contributed articles to a number of other publications.

Sudhindra Sharma is Executive Director of Interdisciplinary Analysts, a research organization based in Kathmandu, Nepal. A sociologist by training, he obtained his PhD from the University of Tampere, Finland, in 2001. He was also awarded the Docent in Development Studies at the University of Helsinki in February 2009.

Sunil Thrikawala is Resource Economist, the Dam Safety and Water Resources Planning Project and Senior Consultant at the Department of Plantation & Engineering, Open University of Sri Lanka. An agricultural economist working in the areas of resource economics, environmental economics and integrated water resources management (IWRM), he has a PhD in IWRM. He has several international and local publications in refereed journals and books.

Index

Accelerated Rural Water Supply Programme Guidelines 11
access: to drinking water 133, 193–4; to recharged water 60
accountability 4, 17, 22, 32–4, 39, 41–2, 44–5, 47, 64, 105, 108–9, 111, 184, 194, 242, 253, *see also* transparency
Act of Parliament (1979), Sri Lanka 240–1
Adarsh Gaon Yojana in Maharashtra 61
Adhav, Baba 293
Adivasi Morcha Sangathan 111
Adivasi Shakti Sangathan 111
Aga Khan Rural Support Programme 79
Agragamee 79
Agricultural: production 70, 164
Air (Prevention and Control of Pollution) Act 1981 311
All India Coordination Committee of Communist Revolutionaries and CPI (Marxist-Leninist) 93
Alternatives and Opposition (IAO) Network 271–2
Andhi Khola Irrigation Scheme in Nepal 5
Anti-Corruption Commission (ACC) 280
anti-dam agitation 70, *see also* Narmada Bachao Andolan
anti-pollution measures 250, 257
Asian Development Bank (ADB) 8, 22, 133–4, 148, 201, 245, 266; in Bangladesh 266, 279–80; funding 8; in Nepal 131–3; policy on integration of water service 215; in Sri Lanka 148, 205, 207, 209, 211, 214–5, 219, 221, 223, 252; Sri Lankan Water policy and 203
Association of Development Agencies in Bangladesh (ADAB) 269
Associazione Internazionale Disarmo e Sviluppo (DISVI), Nepal 133

Baba Amte Centre 110
Bachawat Award of 1975 293
Bagli 91, 98–9, 108–11, 114–15
Baliraja dam movement 299
Bangladesh 3, 5, 8, 19, 22–3, 35–6, 38–44, 46–7, 267–76, 277–80; flood disasters in 267–8, 270; French engineers to 266; J.A. Krug on flood in 264; Monowar Hossain on 264; Prasanta Chandra Mahalanobis on 264; rivers of 266; wetlands in 263
Bangladesh Citizens' Group 269
Bangladesh Water Development Board (BWDB) 269, 279–80
Bangladeshi NGOs 23, 270, 272–4, 277–8
Bayart, J.F. 91
Berlin-based Information, Alternatives and Opposition (IAO) 271–3
Bharatiya Agro Industries Foundation (BAIF) 72
Bhave, Vinoba 93
Bhoodan (voluntary land redistribution) movement 93–4, *see also* Sarvodaya movement

348   Index

Bhutan 35–6
Biochemical Oxygen Demand (BOD) 161
Biological Diversity Act, 2002 311
Bir Dhara Works 130
British NGOs 273
bulk water use 209, 212, 223–5, 228; common law over 222, *see also under* Sri Lanka

Canadian International Development Agency (CIDA) 273, 275
Centre for Applied Systems Analysis in Development (CASAD) 292, 299–301
Centre for Civil Society 15, 286
Chatterjee, P. 91
check dams 61, 70
China, World Bank funded "Three Gorges Dam" in 275
Chittoorpuzha irrigation project 312
Chowdhury, J.U. 269
civil society 1–2, 15–22, 32, 34–5, 38–40, 42–3, 45–8, 55–6, 65–7 76–81, 90–1, 97, 196, 217–18, 261, 289–90, 293; concept of 16, 42–3; Robert Putnam on 43
Civil Society Organizations (CSOs) 1, 17, 20, 22, 32, 34, 38, 43–6, 48, 55–6, 67–71, 73, 80–1, 83, 96, 194, 204, 213, 217, 227; Fred Powell and Geoghegan on 44
Coalition of Environmental NGOs (CEN) 269, 277
Coca-Cola 310–4, 318, 317–9
Coca-Cola Virudha Janakeeya Samara Samithy 313, 319
commodification: of water 203, 228; agricultural water 7, 202
Common Guidelines 57–8, 61, 77, 79, 102
Common Minimum Programme (CMP) 302

community development Organizations 65–6; Jack Rothman on models of 66
Community Water Supply (CWS), Nepal 131–4
community: participation 188, 196; resources 167, 329–30, 336, 338
community-based organisations (CBOs) 43, 60, 78, 95–6, 166
Compartmentalization Pilot Project (CPP) 268–70
Compensation Claims Special Tribunal Bill 310
Comprehensive Water Resources Management Project (CWRMP) 201, 215, 229
conservation of water 61, 100, 103, 225
consultancy 191; overspending in 164
contamination 145, 156, 160, *see also* pollution
Cooperative for American Relief Everywhere (CARE)-Nepal 133
corruption 33–4, 37, 47, 280, 324
Council for Promotion of Participatory Action and Rural Technology (CAPART) 95, 110

dalits 13, 67, 138–9, 319
dams 3, 61, 67, 70–2, 131, 202–3, 294, 295–6, 298, 304; affected people 293–4, 297; beneficiaries 296–7, 300; construction in Krishna valley 294; as temples of development 69
Dandekar, V.M. 291
Dandekar-Deuskar-Deshmukh committee/3D committee 291
Danish International Development Agency (DANIDA) 180–1
Datye, K. R. 290, 299

decentralisation 1, 32–4, 41, 102, 111, 114, 333
Deduru Oya 247
demand side water management 7, 170
demand-responsive approach 186, 194
democratic: participation 1–2; political system 35–6
Dempster, Jim 275
Department of Water Supply and Sewerage (DWSS), Nepal 131–4, 136–8
Deshmukh, Datta 291–3
Deuskar, V. R. 291
development non-governmental organizations. *See* non-governmental organizations
Development Support Centre 79
development: agency 74, 127; assistance 3, 128, 147, 181; and crop destruction 46; interventions 13, 127; NGOs for 66, 68, 81; projects 68, 103, 110, 115, 279
Dharangrasta Parishad 293–5
Dhom dam 305, *see also* dams
domestic water supply 62, 72, 123, 129, 136–7, 139
donor agencies 1, 3, 6, 20–1, 32, 46–7, 57–8, 61, 69, 80, 96, 128, 133, 137–8, 148–9, 159–64, 166, 205–7, 213, 216, 231–2, 269–75, 277–80, 283; agenda of 272; pressure 220–1; in Sri Lanka 167
drainage schemes 3, 265; indigenous design for 158
drinking water 9–10, 173–4, 182, 192
Drought Relief and Rehabilitation Committee, Maharashtra 291
drought-prone areas 301–9
Dublin Principles, and water as "economic good" 203

Dutch Flood Action Plan Platform 274
Dutch government 274, 283
Dutch NGO Flood Action Plan Platform 277

'easementary right' 107
economic reforms 36–7, 47, 68, 75
embankments 3, 131, 267, 269, 276, 286, 324, 331, *see also* check dams; flood control
Employment Guarantee Scheme (EGS) 70, 290, 304
Encountering Development, of Arturo Escobar 125
entitlements system 5, 10; in Sri Lanka 205, 209, 223–9, 233
Environment (Protection) Act, 1986 317, 321
Environmental Impact Assessment (EIA), Sri Lanka 147, 245, 256, 258
equitable water distribution 23, 82, 213, 294, 296, 301, 303
Ershad, Hussain Muhammad 266
Escobar, A. 125, 127, 140
European Flood Action Plan Platform 274

farmer participation 7, 202
Farmers' Organizations (FOs) 225
Ferguson, James 96, 125, 127
Fernandez, Aloysius P. 79, *see also* Mysore Resettlement and Development Agency (MYRADA)
Finnish aid-funded Rural Water Supply and Sanitation Project, Nepal 134
Flood Action Plan Bangladesh (FAP) 3, 8, 22–3, 267–74; grounds for opposing 269–70; opposition to 268

flood control 3, 267, 269–80
Flood Plan Coordination Organisation (FPCO), Bangladesh 267–8
Food and Agriculture Organization (FAO), Sri Lanka 204
Foreign Currency Regulation Act (FCRA) 68, 94
foreign aid/funding 21, 44, 47, 67–8, 80, 91, 94, 110, 113, 139, 147–50, 160, 196, 272–3, (*see also* donor agencies); dependency on 157, 163; in Kandy 165
Forest (Conservation) Act, 1980 311
Forest Rights Act 81
Forum for Watershed Research and Policy Dialogue (ForWaRD) 76, 79, 86n30, 31
Forum of Environmental Journalists in Bangladesh 269
Fourth Rural Water Supply Sector Project, Nepal 134
Friends of River Narmada 111, *see also* Narmada Bachao Andolan (NBA)

Gal Oya Water Management Project, Sri Lanka 202
Gandhi, Indira 94
German Agency for Technical Cooperation, Sri Lanka 154
German NGOs 273
good governance 32–4, 36, 41–2, 48; barrier to 42
Gore, Vilas 299
Governance 34–9, 37–8; A.W. Rhodes on 33; concept of 31–4, 37; people's participation in 42; Rosenau on concept of 33
Government Organised NGO (GONGO) 181, 188, 191–2

gram sabha 108, 311, 325–6, 332–4
Gram Vikas 79
grassroots organisations (GROs) 65, 96, 112
Grassroots Support Organisation (GSO) 96
grassroots workers 111, 113
Green Revolution 3, 69, 71, 73, 265
groundwater 73, 211, 235, 237, 242, 251, 260, 309–15, 317–22, 324, 326, 329, 330–1, 333–4; augmenting 60; availability of 206; Coca-Cola polluting and depleting 310; for commercial purposes 243; data on 206; democratic participatory governance system at 23, 310; economic use of 247; extraction in Sri Lanka 73, 205, 243, 249, 256–7, 316; gram sabha for 311; in Kerala 312; to land rights 207; as private property 60; quality of 207; recharge schemes 187; right to access 10, 207; as rural drinking water 206; safe drinking water 256
Gupta, S. 148

Hanumantha Rao Committee 57, 76
Hazare, Anna 109, *see also* Ralegaon Siddhi
hegemonic ideas 16–17
Heller, P. 147
Hindu rate of growth 36, 50n15
Hindustan Coca-Cola Beverages Private Limited (HCCB) 312–22
Hossain, Monowar 3, 264

Independent Water Resources Regulatory Authority 12

Index ≋ 351

Indian Council of Agricultural Research (ICAR) 57
Indian Easements Act of 1882 107, 331
Indo-German Watershed Development Program 74, 76
industrialization 207, 322; In Sri Lanka 243
Information, Alternatives and Opposition (IAO), Berlin-based 271
Initial Environmental Examination (IEE), Sri Lanka 258
institutional arrangements 7, 59, 62, 135, 166, 205, 210, 212, 215, 219, 220, 226–7, 229, 249–50
integrated water resource management (IWRM) 7, 12, 203, 207, 212, 216–7, 230; definition of 25n8
International Drinking Water Supply and Sanitation Decade 132–3
International River Network (IRN) 273, 276
International Water Management Institution (IWMI) 203, 223; in Sri Lanka's water policy process 223
International Water Tribunal 272, 277
international: consultancy, dependency on 184; financial institutions 1, 3, 7, 20, 146, 170, 194; NGO donors 23, 271; NGOs 94–5, 272, 276
Inter-State Water Disputes Act, 1956 324
Inter-State Water Disputes Tribunal 327
IRA Independent Water Resources Regulatory 12
irrigation law reforms 9, 15
Irrigation Management Policy Support Activity (IMPSA) 202–3; Sri Lanka 203

irrigation projects (large) 3, 9–10, 14, 71; construction of 2; as "hydraulic mission" 69

Jalanidhi 181, 184, 186–93; documents 186; KRWSA on 192; World Bank-aided 181
Janatako Afno Khane Pani Ra Sarsafai (JAKPAS), Nepal 134
Japan International Cooperation Agency (JICA), Sri Lanka 155, 160
Japanese Bank of International Cooperation 21, 145, 158, 163, 183–6, 193; and KWA 184–5, 193

Kaleni Conservation Barrage 205
Kandy City Water Supply Augmentation and Environmental Improvement Project (KCWSAEIP) 145, 157
Kandy Municipal Council (KMC) 146, 149, 152, 154, 162–4
Karnataka Watershed Development Society (KAWAD) 61
Kaviraj, S. 17, 91
Kerala Groundwater (Control and Regulation) Act 316–7, 321
Kerala Rural Water and Sanitation Agency (KRWSA) 181, 184, 186, 188, 189–90; World Bank loan to 188–9; foreign assistance to 193
Kerala State Pollution Control Board (KSPCB) 314, 316–22, 325
Kerala Water Authority (KWA) 172–4, 176–7, 180–4, 189–94; Danish International Development Agency (DANIDA) to 180–1; fund allocation to 179; Japan's Overseas Development

Assistance to 183; Netherlands/ Dutch government aid to 180; as para-state body 178; piped water supply of 192; rural drinking water schemes by 176; water-selling income of 177
Khan, Akbar Ali 44
Khan, Mahbub 100–1, 106–8
Kjaer, Anne Mette 33
Kosi Agreement 131
Krishna river basin 294, 297, 300, 302, 305, 306; water allocations in 310
Krug Mission Water and Flood Report 3, 265

Lal Nishan Party 291, 295, 297
Land 84n8
Landell-Mills, Pierre 276
legitimacy 15, 33, 38, 44, 77, 94, 111, 305, *see also* transparency
liberalisation 19, 82; and privatization of water 203, 206, 226, 248
livelihoods, NGOs and 72–3; sustainability of 56, 60–5, 69, 72, 82
lobbying and networking 81
local communities 5, 10, 21, 23, 59, 64–5, 77, 81, 91, 247; relationship between 65;struggles by 318
Local Water Fund 218
Lok Vidnyan Sanghatana (LVS) 290-2, 294–6
Low External Input-Based Sustainable Agriculture (LEISA) 63
Lutheran World Service, Nepal 133

Maharashtra Krishna Valley Development Corporation (MKVDC) 297
Maharashtra Rajya Dharangrastava Prakalpagrasta Parishad 293
Maharashtra Rajya Dharangrastava Prakalpagrasta Shetkari Parishad 293
Maharashtra Rehabilitation of The Project-Affected Act in 1986 293
Mahaweli Authority of Sri Lanka (MASL) 240–1
Mahaweli Diversion Scheme, Sri Lanka 240
Mahaweli project 241
Maoist movements in Hyderabad 94
Membership Support Organisation (MSO) 96, 117n6
Mihir Shah of Samaj Pragati Sahayog 79
millennium development goals (MDGs) 144
Ministry of Agriculture, Lands, Livestock and Irrigation (MALLI), on traditional approaches and practices 221
Ministry of Mahaweli, River Basin and Rajarata Development (MMRBRD) 219–21
minor water bodies 23, 310, 326, 332–4
Mitterrand, Danielle 270
Mitterrand, Francois 270
mobilization 92, 115; collective 111; community 76; strategies of political 23;
Moolathara irrigation project 312
Morse Report 70
Mukti Sangharsh Movement (MSM) 70, 290–2, 294, 299, 300–1, 306
Mysore Resettlement and Development Agency (MYRADA) 73–4, 77, 79

naala agreement 107–8
Narayan, J. P. 93, 95

Narmada Bachao Andolan (NBA) 70–1, 96
Narmada Valley Project 39, 96–7
National Alliance of People's Movements (NAPM) 70
National Audubon Society (NAS) 273
National Environmental Act, Sri Lanka 245, 256–9
National Green Tribunal Act 2010 311
National Water Fund 218
National Water Policy (NWP) 8, 71, 247, 249, 251, 280, 322
National Water Resources Authority (NWRA), Sri Lanka 204–5, 208, 210, 215, 218, 219–20, 224, 226–7, 249
National Water Resources Management Council, Sri Lanka 250
National Water Resources Management Policy, Sri Lanka 249
National Water Resources Policy and Institutional Arrangement 204, 213, 229
National Water Supply and Drainage Board (NWSDB), Sri Lanka 145–6, 148–9, 151, 154, 156–7, 162, 164, 226, 241–2; schemes of 246
National Watershed Development Programme for Rainfed Areas 57
Naxalbari and Peasant Struggle Assistance Committee (NKSSS) 93
Naxalbari movement 93–4, see also Maoist movements in Hyderabad
Nepal 21, 35, 46, 124–8, 131, 133, 135, 137–40; access to piped water 128; American aid programme 129; Colombo Plan of 129; Community Water Supply and Sanitation programme 134; Development Board Act 131; dhungedhara in 123, 130, 140n1; domestic water in 123; domestic water supply 129; First Rural Water Supply Sector Project 133; as least developed country 128; lending agencies in 131; *Pani Goswara* in 131; rural water supply 128; sources of water in 123; water supply in 138
Nepal Red Cross Society 133, 135
Nepal Water for Health (NEWAH) 135
NGOisation, of the 'development' sector 68, 75, 82
nijbal satyagraha 107
NM Sadguru Water and Development Foundation 79
non-governmental organization (NGOs) 4, 16–19, 21–2, 24, 38–9, 44–8, 58, 65–6, 72–83, 90–2, 94–8, 101–5, 113–17, 133–4, 181, 186–7, 194, 204, 211, 213, 221, 227, 269–74, 286–9, as politically neutral development organisations 96; as public service contractors 18; relationship between state and 80, 114 Swapan Garain on types of 66; watershed projects implemented by 73–4; Willets on 286
non-revenue water 152–3
North American NGOs 273
Northern (Europe and North America) NGO partners 278

organic pollution 161–2, see also contamination
other backward classes (OBCs) 39, 311
Overseas Economic Cooperation Fund (OECF) of Japan *See* Japan Bank for International Cooperation

Overseas Inspection Department 274
oxidation ditch 161

Paddy Lands Act of 1958, Ceyl 239, 241
Pakistan 3, 35–6, 38, 40–1, 43–4, 47; Larry Diamond on 47
Panchayati Raj Act, 1993 314, 334; 73rd constitutional amendment on 78, 331, 334
Panchayati Raj Institutions (PRIs) 6, 78, 330
Panchayats (Extension to the Scheduled Areas) Act, 1996 325
Pani Goswara 130–1
Pani Panchayat 5, 109, 299
Pani Sangharsh Chalwal 70–1
Parambikulam-Aliyar Project (PAP) 312
Paranjape, Suhas 23, 299
Parbatya Chattagram Jana Samhati Samiti 40–1
Parthasarathy Technical Committee 58, 76, 79, 82
"participatory institutions, elite capture" of 77
Participatory Irrigation Management (PIM) 9–10
participatory: rural appraisal 64, 77; watershed development 76
Pawar, Sharad 305
people's participation 33, 38–9, 41–2, 45, 78, 274
People's Science Institute 79, 318
People's Science Movement (MSM) 295
Pigg, S. L. 126–7, 138
piped water 13, 21, 124, 128, 130, 136, 139, 152
Plachimada struggle 309, 318, 320, 329; as "politically motivated" 319

policy development process 8, 203–4, 208, 210, 213, 216, 218, 229; in Sri Lanka 208
policy ownership 211, 229
political reform 7, 202
polluter pays principle 258–9, 328
pollution 5, 22, 67, 147, 155, 157–8, 243, 252, 258, 313, 321, 324, 326, 328; "precautionary principle" to prevent 327
Powell, Fred 44
privatization of water 9, 14, 82, 203, 206, 226, 248, 256; Sri Lanka 254–6
Professional Assistance for Development Action 79
Proshika, Bangladesh 4–5
Public Health Engineering Department (PHED) 173
Public Liability Insurance Act, 1991 311
Public Utilities Commission of Sri Lanka Act (2002) 256
Public-Private Partnerships (PPPs) 9, 19, 82, 171
Purogami Stree Sangathana 94

Rahman, M.M. 271
rainfed regions of India 58, 61
rajakariya system, abolition of 238
Rajiv Gandhi Mission 103
Ralegaon Siddhi 57, 71, 109
reforms 1, 5–11, 13–15, 22, 37, 41, 47–8, 68, 75, 113, 145, 170–1, 202–4, 206, 212–3, 217, 223, 227–8
regulatory agencies 9, 173
rehabilitation 258, 291, 293–6, 298, 329
Report of the Expert Group on Groundwater Management and Ownership 329
right of access to water 322

River Basin Committees (RBCs), Sri Lanka 204, 224, 250
Roman-Dutch law 237, *see also* Sri Lanka, common law in
rule of law 33–4, 327
rural water supply 11, 128, 133–5, 146, 149, 183, 192

safe drinking water 146, 256; in rural areas 44
safe water 11, 21, 207; replacing piped water 124
Samaj Pragati Sahyog (SPS) 79, 98
Sardar Sarovar Project 4, 70, *see also* Narmada Valley Project
Sarvodaya movement 93–4
Scheduled Tribes and Other Traditional Forest Dwellers (Recognition of Forest Rights) Act, The (2006) 333–4
self help groups (SHGs) 16, 74, 76, 286
self-governing rules 33–4, 37
semi-government organisations (SGOs) 44
Seva Mandir 79
sewage treatment plant (STP) 158–9, 162, 165–6
Shetmajoor Kashtakari Shetkari Sanghatana (SKSS) 289, 300, 302–3
Shramik Mukti Dal (SMD) 290, 292
Shumshere, Chandra, expanding piped water supply system 130
Singh, Lakhan 100, 107
Smith, William 276
social action groups 65–7, 69, 95
social movements 16–19, 21, 23–4, 43, 48, 66–7, 69–71, 82, 290, 292–4, 311; against large dams 71; Byrne on 289; Omvedt on 67; Wilson on 288

Social Services National Coordination Committee, Nepal, NGOs affiliated to 133
social: audit 111; capital 43; equity 1–2, 6
socialisms 16–17
Society for Promoting Participative Ecosystem Management (SOPPECOM) 296
socio-religious reform 92, *see also* reforms
South Maharashtra Movement for water rights (SMM) 290–2, 294, 296, 299, 306
Southwest Area Integrated Water Resource Management 279
Sri Lanka 7, 12, 21–2, 38, 145, 148–9, 156, 165, 201–2, 203, 224, 236, 238, 243–4, 253, 257, 259; anti-pollution measures 257; Asian Development Bank (ADB) in 22; bethma system in 238; commodification and privatization of water resources in 206; common law in 222, 237; donors for adoption of "water rights" in 203; Eppawela" and "Thuruwila" in 244; farmer participation in 202; Foreign Aid for NWSDB in 155; funded by the Japan Bank for International Cooperation 145; grants and loans to 149; groundwater rights in 207; Mahaweli in 157; National Environmental Act 245; National Water Supply and Drainage Board (NWSDB) in 145–6; piped water in 151; policy development process in 8; pollution in 157; rights to water in 5; safe drinking water in 145; small water users 205, 209, 211, 223–4; social movement in 21; Thuruwila 249, 255, 263;

waste water disposal in 146; water policy in 205, 252; water resources management in 205; water resources of 217, 251; water sector reforms in 145; water supply schemes 250
State-Civil Society Relations 45–7, 92–7
State Groundwater Authorities 331, 333
State Pollution Control Board 337
Stree Mukti Sanghatana 94
Sukhomajri 5, 57, 71
Support for Social Progress (SPS) 98–116; in Bagli 109, 116; and *naala* agreement 107; Neelpura solidarity with 108; opposition to 102
Support for Social Progress (SPS) in NGO-state relations 113
surface water 60, 206
sustainable water resources management 22, 214
Swiss NGO Helvetas, Nepal 132

Takari Lift Irrigation Scheme 300
Takari struggle 301
Tamil Nadu–Kerala interstate project 312
technology for waste water treatment 161
technology transfer 3
Telkamper, Wilfried 274, 276
Tembu Lift Irrigation Scheme 302
Tembu scheme 303; for equitable water distribution 303
Third Rural Water Supply Sector Project, Nepal 134
thiyya andolans 183, *see also* Narmada Bachao Andolan
Tokyo Engineering Company (TEC) 183; technical expertise of 297
tradable rights 229; donor agencies for 209

Trade Related Intellectual Property Rights (TRIPS) Agreement, Sri Lanka 253
traditional village water governance system, marginalization of 327
transferable water entitlements, Sri Lanka 12, 248
transparency 4, 17, 22, 32–4, 38–9, 41, 44–7, 105, 111, 116, 184, 191, 194, 230, 242, 279
Transparency International Bangladesh (TIB) 47
treating/reusing water 165

Uchangi dam, Kolhapur 296
Ujani dams 305
United Nations Children's Fund (UNICEF), in Nepal 133
United Nations Development Programme (UNDP) 133, 266, 276
United States Agency for International Development (USAID) 3, 7, 32, 34, 201–2, 265
universal access to drinking water 133
urban water supply: and sanitation 134, 166, 193, 241; schemes in Kerala 146, 182, 193
urbanization 157, 207
Urmodi dam 295, 305

Varshney, Ashutosh 39
velvidana (irrigation headman) system 239
Vidnyan Jatra 1984 291
voluntary organizations (VOs) 43–4, 46, 92, 94–5, 171, 287

Wasteland Integrated Research Program (WIRP) 292, 300
Water (Prevention and Control of Pollution) Act (1974) 311, 317, 321, 324

Water Aid and Redd Barna, Nepal 133
water and sanitation services 20–1, 135; Sri Lanka 123, 137, 147, 149–50
Water Management Improvement Project (WMIP) 279
water policies 135, 137, 213–4, 219, 224, 248
Water Policy Development 22, 205, 207, 209, 211, 213, 215, 217, 219, 221–3
Water Pollution Prevention and Control Acts 334
Water Resources Act 204, 220, 224, 226; Sri Lanka 228
Water Resources Authority Act, Sri Lanka 252
Water Resources Board Act (1964) Sri Lanka 224
Water Resources Council (WRC), Sri Lanka 204, 210, 219, 220
water resources management 1, 205, 208–10, 212, 219, 222, 224, 229, 232, 253, 269, 274, 326; Sri Lanka's perception of 207
Water Resources Management Project (WRMP) 205, 214–7, 219
Water Resources Management Tribunal, Sri Lanka 250–1
Water Resources Master Plan, Sri Lanka 207
Water Resources Secretariat (WRS), Sri Lanka 204
Water Resources Tribunal (WRT), Sri Lanka 219–20, 250
Water Services Reforms Bill, Sri Lanka 226
Water Users' Associations (WUAs) 8–10, 14–15, 305; water distribution among of members 10
water: decentralization of responsibilities over 327; as economic good 13, 192, 194; market 183, 322; policy in India 25n6; resource allocation 4, 215; rights 5, 207, 210, 212, 222–4, 227, 236, 238, 245, 250, 253, 257, 260, 290, 293, 307, 318, 321; scarcity 7, 183, 322; sector reforms 1, 6, 12, 144; sewerage systems in Europe 141n18; as state subject 3; storage 3, 10, 324, 331; supply schemes 183, 187, 190, 207, 242, 331, 333; tariff in Kerala 14, 150, 179; users of 8, 14, 206, 212–13, 222–3, 230, 331
water-borne diseases 124, 158
water-intensive crops 12, 248, 292
water-intensive urbanization 326
watershed development 5, 20, 55–65, 68–82, 102–5, 110, 112, 116, 292, 325, 331; civil society action in 65–9, 78–83; Common Principles for 61; measures of 60; for Neelpura 103; as participatory 76–8; policies of 20, 55–8
Watershed Organisation Trust and Hind Swaraj Trust 79
Watershed Support Services and Activities Network (WASSAN, Hyderabad) 76, 79
Willcocks, William 265
World Bank 3–4, 7, 32–3, 69–70, 128, 131–2, 134, 149, 173, 176, 180, 181–2, 184, 186, 202, 266–70, 273, 278, 280; Government borrowing from 186
World Bank-funded Water Supply and Sewerage Project, in Kathmandu 132
World Trade Organization (WTO) 17, *see also* liberalization
world water market 223
www.jalanidhi.org 186